Key Concepts in Physics

Key Concepts in Physics

Nelson Bolivar

ARCLER
PRESS

www.arclerpress.com

Key Concepts in Physics

Nelson Bolivar

Arcler Press
224 Shoreacres Road
Burlington, ON L7L 2H2
Canada
www.arclerpress.com
Email: orders@arclereducation.com

ISBN: 978-1-77469-149-6 (Hardcover)

ABOUT THE AUTHOR

Nelson Bolivar is currently a Physics Professor in the Physics Department at the Universidad Central de Venezuela, where he has been teaching since 2007. His interests include quantum field theory applied in condensed matter. He obtained his PhD in physics from the Universite de Lorraine (France) in 2014 in a joint PhD with the Universidad Central de Venezuela. His BSc in physics is from the Universidad Central de Venezuela.

TABLE OF CONTENTS

LIST OF FIGURES

LIST OF ABBREVIATIONS

CGS	Centimeter–Gram–Second
JFET	Junction-Gate Field-Effect Transistor
IGFET	Insulated-Gate Field-Effect Transistor
MOSFET	Metal Oxide Semiconductor Field-Effect Transistor
DIODE	Dynamic Information Operations Decision Environment
RAM	Random Access Memory
ROM	Read-Only Memory
TNT	Trinitrotoluene

PREFACE

Physics is a branch of science that deals with the structure of matter as well as the interchange between the basic components of the universe that examines everything. In an extensive manner, physics is derived from the Greek word physikos which includes concepts and dimensions of nature on both the levels, be it macroscopic or submicroscopic.

It's a range of study which revolves around not only the behavior of objects under the action of provided forces but is also concerned with the concept and origin of different types of forces such as gravitational, electromagnetic, and nuclear force. Its utmost goal is the construction of a few far-reaching principles that draw together and describe all such various occurring phenomena.

Physics is the fundamental physical science. Until rather during current years, physics, in addition to the natural philosophy, has been applied conversely for the science whose objective is the discovery and construction of the basic laws of nature. As the contemporary sciences evolved and became expandingly specialized, physics came to imply that a fraction of physical science is excluded from astronomy, chemistry, geology, as well as engineering.

Physics plays a vital role in all the natural sciences, although, and all such fields contains branches where physical laws and systematic evaluations receive an exceptional emphasis, bearing names like astrophysics, geophysics, biophysics, along with psychophysics. Physics can, at the initial stage, be explained as the science of matter, motion, and energy. Its laws are generally defined with economy and precision in the field and language of mathematics.

Both experiments, the examination of phenomena under circumstances that are regulated as accurately as possible, and theory, the formation of a combined conceptual structure, play a significant and harmonizing roles in the progression of physics.

Physical experiments result in estimations, which are contrasted with the aftermath anticipated by theory. A theory that reliably anticipates the outcome of experiments to which it is relevant is said to incorporated a law of physics. Although, a law is always subject to mitigation, replacement, or curtailment to a more restricted domain, if eventually the experiment makes it required.

The major focus and objective of physics is to discover a unified framework of laws governing matter, motion, as well as energy at small (microscopic) subatomic distances, at the human extensive (macroscopic) scale of day-to-day life, and out to the largest distances (for instance, those on the extragalactic scale). This aspiring goal has been identified to a remarkable extent. However, an entire unified theory of physical phenomena has not yet been obtained (and plausibly never will be), a notable small set of basic physical laws emerges to be accountable for all known phenomena.

The contemporary developments of relativity as well as quantum mechanics modify these laws insofar as they implement to higher speeds, very immense objects, and to the tiny elementary components of matter, such as electrons, protons, along with neutrons.

This book will introduce the readers to the field of physics and its concepts. It is precisely designed for readers or students with no prior experience related to physics and its key concepts, and touches upon a diversity of fundamental topics. By the end of the book, readers will understand the basic knowledge of dimensions of physics and the key concepts involved.

A

Aberration

- An aberration is a flaw in an optical system that causes visual distortion. Astigmatism, Chromatic Aberration, Coma, Curvature of Field, Optical Distortion, and Spherical Aberration are some of the different types of aberration.
- The apparent shift in the position of a star caused by the Earth's orbital motion.

Abrasive

A material used in polishing or cleaning to wear away a solid surface. Emery and pumice are two common examples.

Absicissa

A point's x co-ordinate on a two-dimensional Cartesian graph, i.e., its distance from the y axis, measured along the x axis from the origin.

Absolute Pressure

The real pressure at a given location as a result of all factors. The total of air pressure and pressure due to the supernatant liquid determines the absolute pressure at a point in a liquid.

Absorption

All or a portion of the energy incident on a substance is converted into another type of energy within the substance; for example, part of the energy of an incident beam of light may be used to excite the atoms or molecules of the absorbing substance.

Absorption Coefficient

Symbol α. is a quantity defined by the equation $I = I_0e^{-\alpha x}$, where I_0 and I are the incident and transmitted radiation intensities, respectively, and x is the thickness of the material traversed. The quantity x can be written as a length, mass per unit area, moles per unit area, or atoms per unit area; the associated absorption coefficients are linear, mass, molar, and atomic, respectively.

Absorption Spectrum

A SPECTRUM produced by the absorption of electromagnetic radiation by matter. To obtain an absorption spectrum, incident radiation with a continuous range of wavelengths is used.

The prism or diffraction grating is used to disperse the beam emerging from the sample. It is discovered that some wavelengths of radiation have been absorbed, and the energy at these wavelengths has been used to raise the energy of atoms or molecules in the absorbing medium to higher levels. When visible light is incident, the absorption spectrum is made up of dark lines or bands on a bright continuous background.

Abundance

- Symbol C. The ratio of a specific ISOTOPE's number of atoms to the total number of atoms in an isotope mixture. Natural abundance refers to the amount of an element's isotopic mixture that occurs naturally.
- The amount of a particular substance in the Earth's crust (also known as terrestrial abundance) or the universe (sometimes called cosmic abundance).

Acceleration Due To Gravity

Symbol g. The acceleration of a body falling freely, without air resistance or buoyancy, at a particular point close to the Earth's surface. It is caused by the gravitational attraction of the Earth and varies with latitude because the Earth is not completely spherical and also because of the Earth's rotation. At the poles g is 9.832 metre per second^2 (m s^{-2}); at the equator it is 9.78 m s^{-2}; the Earth's rotation accounts for 0.034 m s^{-2} of the difference. The standard value of g (symbol gn) is taken as 9.806 65 m s^{-2}.

Accelerator

The kinetic energy of charged ions and ELEMENTARY PARTICLES can be increased with this equipment.

Its primary applications are in nuclear and particle physics fundamental research as well as radiotherapy. Betatron; Cockcroft-Walton Generator; Cyclotron; Electrostatic Generator; Linear Accelerator; Synchrocyclotron; Synchrotron; Van De Graaff Generator.

Acceptor

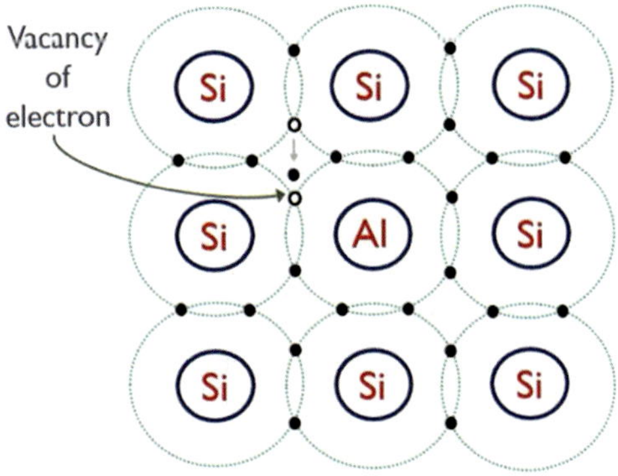

Figure 1. Acceptor.

Source: Image by Wikimedia commons.

An atom that receives an electron in an external SEMICONDUCTOR, resulting in a positive HOLE in the conduction band.

Access Time

The amount of time it takes for information to be obtained from a computer store and processed.

Accumulator

A group of secondary cells that are connected in a series. The charge on an accumulator is the entire quantity of electricity stored; it is commonly measured in ampere-hour.

Accuracy

The degree to which an experimentally determined value and the true value agree. Operator faults, instrument errors, and random errors all have an impact. The first can be eliminated with care and patience; the second is inherent: the limitations of any instrument should always be addressed; and the third can be avoided by using as large a numbing agent as practicable.

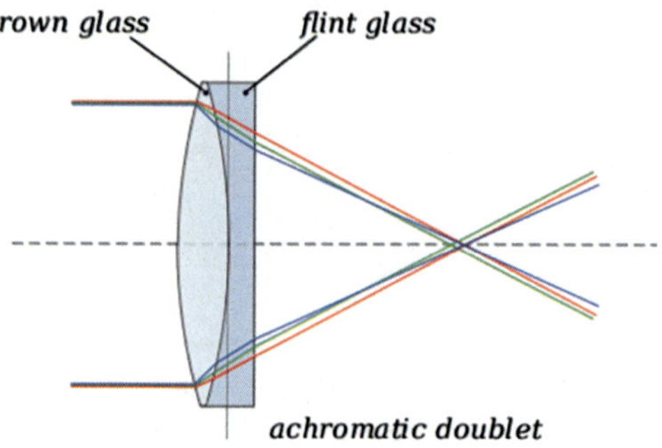

Figure 2. Accuracy.

Source: Image by Wikimedia commons.

Achromatic Lens

A planoconvex combination lens made up of a biconvex crown glass lens bonded to a planoconcave flint glass lens by Canada balsam. It is required to image red and blue rays from the same axial object point at the same axial image point.

Achromatic Prism

A compound prism that removes DISPERSION between two colors, most often red and blue. The remaining colors in the light are dispersed in incident white light, but most of the color effect is lost because blue and red are at the two ends of the visible spectrum. Two small angle prisms, one made of crown glass and the other of flint glass, are in contact.

The condition for no red-blue dispersion by the combination (i.e., for blue and red rays to emerge parallel) is (ncB – nCR)AC = (nFB – nFR)AF, where A and n denote prism angle and refractive index, suffixes F and C denote flint and crown glass, and suffixes B and R denote blue and red light, respectively.

Acoustic Measurements

The rate of energy transmission, absorption and dissipation in the medium, and the SPEED OF SOUND in it are all determined by varying the pressure, density, temperature, and particle location of a medium transmitting sound.

Acoustics

- The study of SOUND. Small-amplitude noises, or those below the pain threshold, are the focus of linear acoustics. Large-amplitude sounds, such as those produced by explosives and shock waves, are studied in nonlinear acoustics. Physiological acoustics is the scientific study of the ear's reception of sound and the nervous system's processing of the associated signals.
- The qualities of a room or building that determine how well music and other sounds can be heard within it.

Acoustoelectric Effect

When an acoustic wave passes through a medium containing mobile electric charges, an electrostatic field appears.

Actinometry

- A measurement of how much light a system absorbs during a photochemical transition.
- The measuring of solar radiation's energy.

Action Reaction Pair

According to the third of NEWTON'S LAWS OF MOTION, the equal and opposing forces that bodies exert on one another.

Activate

- To provide an atom or molecule with enough energy to make it reactive.
- To make a material radioactive by bombarding it with neutrons, for example.

Activation Analysis

The identification and quantification of radioactive isotopes produced by irradiating a sample in a nuclear reactor is a method of chemical analysis. The procedure is extremely delicate.

Activation Energy

The amount of energy necessary by an atomic or molecular system to allow a specific operation to take place, such as the energy required to raise an electron to the conduction band or the energy required by a molecule to participate in a chemical reaction.

Additive Process

A procedure in which a colored light is created by combining a large number of different colored lights in a direct color.

Adiabatic Curve

For a given mass of gas conducting an ADIABATIC PROCESS, a plot of pressure against volume is shown. The curve for 1 kilogram of air, as well as an ISOTHERMAL for 273 K and one for 473 K for comparison. The adiabatic through a point is always steeper than the isothermal, as can be observed.

Adiabatic Demagnetization

A method for producing near-absolute zero temperatures by (a) demagnetizing a paramagnetic salt: the magnetized salt is first cooled to liquid helium temperature and then thermally isolated from its surroundings; when the magnetizing field is removed, the sample becomes demagnetized and cools even more.

It has been possible to get temperatures as low as IQ–3 K. (b) adiabatically demagnetizing a substance containing nuclear magnetic moments: the nuclear spins are magnetically aligned isothermally at around 10–2 K, then adiabatically demagnetized. Low temperatures of IQ–6 k are expected.

Adiabatic Process

A method in which no heat is introduced or removed from the system. Normally, the temperature fluctuates. When a gas is compressed adiabatically, for example, work is done on the system and the gas temperature rises. ISOTHERMAL PROCESS is similar to ISOTHERMAL PROCESS.

Adsorption

On a solid surface, the creation of a layer of material, usually gas. ABSORPTION is the polar opposite of ABSORPTION. The solid has not been penetrated. The layer is one atom or one molecule thick if it is bound together by covalent bonds, and the process is known as chemisorption. The process is known as physisorption when the binding is via van der Waals forces and the layer has a thickness of several atoms or molecules.

Aerial

A conductor, or a series of conductors, made up of wire in most cases. It is situated at a higher elevation. An aerial's purpose is to transmit or receive radio waves: in transmission, accelerating charges in the aerial produce electromagnetic waves; in reception, electromagnetic radiation produces minor changing currents in the aerial. There are a variety of aerials available, each suited to a specific frequency range. At frequencies below roughly 30 megahertz, the dipole aerial, for example, is utilized. It's a straight conductor with a length of half a wavelength, but it can also be a full wavelength. At the center point, a connection is created.

Aerial Array

A system comprising two or more AERIALS that are oriented in such a way that they have specific directional radiating or receiving properties.

Aerodynamics

The study of gases in motion, particularly the relative motion of solid objects and gases, such as body motion in air.

Aerofoil

A body that experiences a lift force much greater than the drag force when moving through a fluid. There is a velocity differential between the fluid layers reaching the trailing edge B via the upper and lower surface paths due to viscosity; this causes a vortex at the trailing edge, as well as counter circulation around the aerofoil, which creates the lift force. The usage of aerofoils for wing and tail components is essential to an airplane's flying.

Aerosol

Aerosol is a tiny dispersion of liquid or solid particles in a gas with particle sizes ranging from 1 nanometer to 50 micrometers. Hazes, mists, fogs, clouds, smokes, dusts, live germs, viruses, and molds are all examples. Aerosols play a role in the formation of clouds, precipitation processes, atmospheric chemistry, air pollution, visibility, and radiation transfer in the atmosphere.

Commercial aerosols are formed by dispersing liquids or solids into the atmosphere using a gas, most often chlorofluoromethane, under pressure. The increasing usage of chlorofluoromethane has raised concerns about its buildup in the stratosphere, which could result in ozone depletion. More solar UV light will reach the Earth's surface as a result of this.

Agonic Line

A line on the Earth's surface connecting places of zero magnetic declination, or true north as indicated by a compass. The two main agonic lines are one that runs across America and the other that runs through Eastern Europe, Arabia, Asia, and Australia in an erratic pattern.

Air Breakdown

A CORONA DISCHARGE is a phenomena that happens when the electric field intensity is greatest at a spot on a charged conductor.

Air Cell Method

A method of determining a liquid's refractive index in relation to air. The cell is made up of two thin plane parallel glass plates cemented together to hold a constant thickness film of air. A glass vessel with thin planar parallel sides holds the liquid under inquiry. Light from a monochromatic external source flows through the container and into the liquid, where it generally strikes the submerged air cell. The cell is then rotated around a vertical axis until no light passes through it, or until it is completely internally reflected. On the scale supplied, the angle of rotation, 9, is measured; it can be seen that $1/\sin\theta$ is the required refractive index.

Air Wedge Fringes

An example of light wave INTERFERENCE. A monochromatic light source illuminates a thin wedge of air, and the light reflected at the upper and lower glass slides that confine the wedge is focused by the eye. The FRINGES are evenly spaced and parallel to the wedge edge; if N bright bands are separated by a distance of /, the wedge angle is $N\lambda$ (2l), where, λ is the wavelength of the light used.

Airy Disc

The image of a point source of light that would be created by an optical system with no ABERRATION; the disc is a DIFFRACTION pattern that shrinks in size as the optical system's aperture expands.

Alcohol Thermometer

A type of thermometer that is identical in construction to the MERCURY IN GLASS THERMOMETER, except that the mercury is replaced by alcohol, which is frequently colored for easier viewing. The alcohol thermometer is much more useful at low temperatures than the mercury thermometer because alcohol freezes at –130°C compared to –39°C for mercury's freezing point. Alcohol boils at 78°C, so its upper range is limited compared to mercury's boiling point of 459°C.

Alfven Waves

Under specific conditions, magnetohydrodynamic waves propagated through a PLASMA. Waves flow in the same direction as the applied magnetic field, while plasma particles oscillate in the opposite direction.

Algebra

Variable quantities and numbers are represented by symbols in this discipline of mathematics. Equations are used to define relationships between symbols, which are then structured into handy forms for solution using a set of logical rules.

Algebraic Summation

The result of adding a group of numbers while taking into account their sign; for example, the algebraic sum of 15 and –10 is 5.

Algorithm

A specified set of procedures that, in theory, will result in the solution to a given issue. The sequence can be as long as you want it to be.

Alpha Decay

Radium–226, atomic number 88, decays to radon–222, atomic number 86, by ejecting an ALPHA PARTICLE from an unstable nucleus, generating a nucleus with a MASS NUMBER 4 less than the parent's and an ATOMICNUMBER 2 less than the parent's.

Alpha Ray Spectrometry

The distribution of KINETIC ENERGY or MOMENTUM of alpha particles from a specific radioactive nuclide is measured. The nuclide usually emits a sequence of single-energy alpha particle groups, one of which is dominant. As a result, the alpha ray spectrum has 'lines' that reflect this energy distribution.

Alpha Wave

In ELECTROENCEPHALOGRAPHY, an excellent feature was obtained. The wave frequency is 5 to 13 hertz, with the lowest frequency when the eyes are closed and the maximum frequency when the eyes are open; the wave shape is regular sinusoidal. The amplitude of the alpha wave is greatest near the occipital area, which is linked to vision. The alpha wave is suppressed at first by focusing the eyes, but it reappears later. When you fall asleep, the wave goes away and is replaced by a wave with a frequency of roughly 1 hertz; any stimulation during sleep causes an alpha wave train.

Alternating Current

An electric current that reverses direction on a regular basis. The most common variety has a sinusoidal waveform, which is given by the equation $I = I_0 \sin 2\pi ft$, where I is the current at time t, I_0 is the maximum current value, and f is the alternation frequency. Every period, the alternating current cycles through its cycle of values once, with the average value being zero.

Altitude

- The vertical distance above sea level of a point.
- A geometric or solid's height measured perpendicular to its base.
- The angular distance of a celestial body above the horizon as specified by a celestial co-ordinate.

Ammeter

An electric current meter is a device that measures the flow of electricity. moving coil instrument; moving iron instrument; hot wire ammeter are examples of moving coil instruments.

Figure 3. Ammeter.

Source: Image by Wikimedia commons.

Amount of Substance

Symbol n. The AVOGADRO CONSTANT multiplied by the number of entities that make up the sample under examination; for example, the amount of substance in a sample of light is the number of photons multiplied by the Avogadro constant. The MOLE is the unit.

Ampere

The SI UNITS system's unit of electric current is designated by the letter A. It is defined as the constant current that, when conducted in vacuo through two parallel conductors of infinite length and negligible cross section, creates a force between them of 2×10^{-7} newton per meter of length. This unit replaced the international ampere, which was defined as the current that deposited silver at a rate of 0.001118 grams per second when passed through silver nitrate solution under certain conditions; one international ampere equals 0.999850 ampere.

Ampere Balance

An instrument for measuring absolute current. The zero screw is adjusted so that plane ABCD is horizontal when the current is turned off. The current to be measured is then turned on, and the rheostat is adjusted (if necessary) to maintain a consistent meter reading during the experiment. Because the current flows in a series through ABCD and EFGH, FG repels CB.

The required mass, m kilogram, is then determined; because the distances between CB and the scale pan from the pivot are equal, the force between the conductors is mg newton, where g is the acceleration due to gravity. The spacing (r meter) and equal lengths (l meter) of FG and CB are measured. $[mgr/(2 \times 10^{-7} l)^{1/2}]$ returns the current I ampere. This method is absolute because it is based on the definition of the ampere.

Amplifier

A device that draws energy from a source apart from the signal to boost the intensity of an electric signal. Amplifiers are classed according to the frequency range in which they are intended to operate, such as audio frequency. frequency of radio. frequency at an extremely high level. They can also be categorized based on the type of transistor or valve used.

Amplitude

The maximum numerical value of an alternating quantity.

The angular distance of a celestial body, measured on the horizon, from the east or west point.

The quantity $+ (a^2 + b^2)^{1/2}$ for the complex number (a + ib).

Anastigmat

A three- or four-component camera lens that exhibits minimal ASTIGMATISM, CHROMATIC ABERRATION, COMA, IMAGE CURVATURE, and SPHERICAL ABBERATION even at big apertures and wide fields.

Anelasticity

In the pre-plastic range, a feature of a solid that prevents a unique relationship between STRESS and STRAIN.

Anemometer

A device for determining the velocity of a fluid, particularly wind. The wind's momentum is employed to spin a collection of miniature cups fixed on a spindle; the instrument is calibrated to display the wind speed directly on a dial. HOT WIRE ANEMOMETER; PITOT TIJBE; VENTURI METER are other terms for the same thing.

Angle of Deviation

The figure created by two intersecting lines; the amount of rotation between the lines is measured in degrees or radians, with 180° equaling n radian: a complete turn is thus 2n radian. The angle between two plane surfaces is the angle formed by the perpendiculars in each plane intersecting at a point on the planes.

Angle of Contact

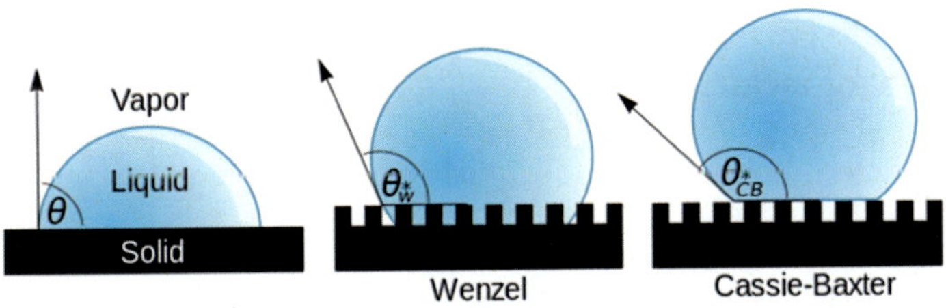

Figure 4. Angle of contact.

Source: Image by Wikimedia commons.

The angle between the wall of a vertical capillary tube and the meniscus of a liquid in the tube, measured in liquid; if the angle is acute, the liquid rises in the tube; if the angle is obtuse, the liquid falls.

Angle of Deviation

The angle at which an incident ray deviates due to interaction with various media, such as reflection or refraction.

Angle of Dynamic Friction

The angle whose tangent is the dynamic FRICTION coefficient. It's the angle of an inclined plane to the horizontal at which a body will glide down without accelerating.

Angle of Incidence

The angle between the point of incidence and the surface; the point of incidence is where the ray hits the surface. The angle formed by a wavefront and the surface it hits.

The Reflection's Angle

At the point of departing, the angle between a ray leaving a reflecting surface and the line perpendicular to the surface (i.e., the normal); this point coincides with the point of incidence. (2) The angle formed by a wavefront and the reflecting surface it leaves behind.

Angle of Refraction

At the point of entrance, the angle formed by a refracted ray and a line perpendicular to the refracting surface (i.e., the normal); this point coincides with the point of incidence.

Angstrom

A is a symbol. The nanometer, which is equivalent to 10 A, has replaced the JO-lo meter as the unit of length. It was previously used for wavelength and interatomic distance measurements.

Annihilation

An annihilation collision occurs when a particle collides with its antiparticle, leading in the conversion of the particles into electromagnetic radiation. When an electron and a positron collide, two gamma ray photons are produced. According to the CONSERVATIO LAW of mass-energy, the total energy of the electromagnetic radiation produced is equal to the sum of the masses and kinetic energies of the particles.

Anomalous Dispersion

In the vicinity of an absorption band, a discontinuity in the regular change of refractive index with DISPERSION. The refractive index of the absorption band increases on the long wavelength side and decreases on the short wavelength side.

Anomalous Expansion of Water

The volume of a particular mass of ice shrinks when it melts, and the volume of the resulting water shrinks even more as the temperature rises up to 4°C. HOPE'S EXPERIMENT can also be found here.

Anomalous Viscosity

VISCOSITY for which Newton's viscosity law is disregarded. All fluids containing two or more phases at the same time, such as every COLLOIDAL SOLUTION, exhibit abnormal viscosity.

Anomaly

An angle that defines a planet's location in its elliptical orbit around the Sun. The genuine anomaly is the angle formed at the Sun by the line connecting the planet with PERIHELION. The mean anomaly is the angle formed at the Sun by a line connecting the perihelion to an imagined planet travelling at constant speed in the real planet's orbit and having the same period as the real planet. The diameter of the semicircle is the orbit's major axis.

Anti – Ferromagnetism

Some inorganic substances, such as manganese oxide, have this feature. It's the result of adjacent magnetic dipole moments being arranged in an antiparallel pattern. Antiferromagnetic materials have a low positive MAGNETIC susceptibility that exhibits a temperature dependence comparable to FERROMAGNETISM. The susceptibility increases with temperature up to the Neel temperature, then drops with temperature according to the CURIE WEISS LAW, indicating that the substance is paramagnetic above the Neel temperature.

Antimatter

The atoms of this hypothetical material are made up of ANTIPARTICLES. Antimatter atoms would have negatively charged nuclei composed of antiprotons and antineutrons, with antielectrons, or positrons, orbiting the nucleus. ANNIHILATION would occur if regular matter and antimatter came into contact. In the universe, antimatter has yet to be discovered.

Antiparticle

An elementary particle with the same mass as a given elementary particle and the same magnitude but opposite sign of CHARGE, BARYON NUMBER, ISOSPIN quantum number, and STRANGENESS as the provided particle.

Aperture Synthesis

Computer synthesis of the outputs of more than one small adjustable RADIO TELESCOPE for varied placements of the telescopes is a means of correctly mapping radio emitters in deep space.

Apparent Change of Weight

- When the interaction between a person and the support is adjusted, the difference in weight sensation is perceived. Thus, in an orbiting space station, the reaction disappears, giving the illusion of weightlessness; in a rocket taking off, the reaction develops to the point where, when lying down, the sensation of considerably increased weight is felt, causing difficulties breathing and raising the arms or legs.
- An upthrust on a body that is completely or partially submerged in a fluid. A body weighed in air and then in a denser medium like water, for example, appears to lose weight. ARCHIMESDES' PRINCIPLE is also worth mentioning.

Apparent Depth

When the medium in which an object is immersed changes, the object's apparent depth changes. As a result, when a swimming pool is depleted of water, the depth appears to increase. The effect is caused by light REFRACTION as it passes through a transparent substance.

- A bright electrical gas discharge with a modest potential gradient and a high current density. The discharge heats the electrodes, and their evaporation aids in maintaining the temperature.
- A curve's open portion.
- The inverse of a circular or hyperbolic function; for example, the arc sine, abbreviated arc sin or sin–1, of a number is any angle whose sine is the same as the number's sine.

Armature

Any moving part in an electrical machine that is induced by a magnetic field to produce a voltage. Rotating coils in an ELECTRIC MOTOR and a ferromagnetic bar attracted by an electromagnet in a relay are two examples. A different term for the artificial pacemaker KEEPER; A device that allows electric impulses to be administered to a heart in order to keep it operating normally. The first pacemakers were implanted in the patient, and the mercury cells that charged up the capacitor whose discharge triggered the heart had to be replaced every 18 months to 2 years. The DEMAND PACEMAKER was the next advancement, followed by the EXTERNAL PACEMAKER.

Aspherical Surface

An aspherical surface is one that is not part of a sphere; it is frequently a part of an ellipsoid, a hyperboloid, or a paraboloid. When compared to lenses and mirrors with spherical surfaces, aspherical surfaces display less aberration, especially Spherical Aberration.

Astatic Galvanometer

Astatic galvanometer is a type of galvanometer in which two comparable magnets are suspended in a sensitive form of moving-magnet galvanometer. Because the couples on each magnet are roughly equal but opposite, the net couple due to an external magnetic field is modest; the current through the coil encircling one of the magnets thus creates a higher rotation for a given current than if only a single magnet were utilized. In a different configuration, each magnet is encircled by a coil, with the two coils twisted in opposite directions to achieve even greater sensitivity.

Asteroid

A tiny planet with a diameter of less than 1000 kilometers. Between the orbits of MARs and JUPITER, most asteroids have near-circular orbits. Small asteroids, on the other hand, have extremely elongated orbits and approach the Sun than do the majority.

Astigmatic Lens

A lens that is used to correct ocular ASTIGMATISM. Depending on the flaw, it can be Plano-cylindrical, sphero-cylindrical, or toroidal.

Astigmatism

An optical aberration in which a point off the system's axis is viewed as two perpendicular lines, the tangential (horizontal) one being closer to the system than the sagittal (vertical).

Astronomy

The study of the universe outside of the atmosphere of the Earth. Astrometry, Astrophysics, and Celestial Mechanics are subsets of the field.

Astrophysics

Astrophysics is the study of the physical processes that occur in celestial bodies and the space between them. The link between star energy and evolution is a major source of worry.

Atmospheric Electricity

The atmosphere's general electrical properties. The following are the mean fine weather values of various characteristics at about sea level:

air-earth current density	2×10^{-14} A m^{-2}
direction of field	downwards
Potential gradient	130 Vm^{-1}
small ion mobility	1.4×10^{-4} $m^2 y^{-1} s^{-1}$
total conductivity	3×10^{-4} S m^{-1}

Ionization by radioactive substances and cosmic rays causes charges to form in the Earth's atmosphere. The electrical characteristics of the atmosphere are significantly altered during a LIGHTNING flash.

Atmospheric Layers

According to the direction of temperature variation, the Earth's atmosphere is separated into gaseous layers. The layers' vertical extents vary over the Earth's surface and reflect seasonal and diurnal fluctuations in the same location. The estimated altitude limitations of each layer as well as temperature variance within the levels. The MESOPAUSE has the lowest temperature.

Atmospheric Optics

The study of light scattering, reflection, refraction, and diffraction by the atmosphere or by ATMOSPHERIC POLLUTION.

Atmospheric Pollution

The presence of undesirable things in the atmosphere, such as dust and poisonous fumes. Pollution like this can obstruct vision, injure plants and animals, and degrade structures.

Atmospheric Pressure

The air above the Earth's surface exerts pressure at sea level.

Atmospherics

Natural electrical atmospheric disturbances cause short-wave trains of electromagnetic radiation. Radio reception is hampered as a result of the radiation.

Atmospheric Windows

Some space radiation penetrates the Earth's atmosphere through gaps in atmospheric absorption. The optical window allows wavelengths between 300 and 900 nanometers, the radio window 8 millimeters to 20 meters, and many small infrared windows at micrometer wavelengths.

Atom

The tiniest part of an element capable of participating in a chemical reaction. Originally, the atom was assumed to be indivisible, but later research led to the concept of the atom as a micro planet, with a Nucleus made up of a precise number of protons and neutrons at its core and a number of electrons surrounding it. An atom's diameter, or atomic diameter, is on the order of l to 10 meters. The nucleus is on the order of 10–15 meter in size.

Geiger and Marsden's Experiment; Bohr Theory; Chadwick's Experiment. The pattern of electron energy levels is given by four quantum numbers: the azimuthal quantum number, the magnetic quantum number, the principal quantum number, and the spin. *Wave Mechanics* and *Quantum Mechanics* introduced further refinements to the conception of the atom.

B

Back Electromotive Force

An electromotive force opposing the flow of current in a circuit.

- of induction: the back electromotive force which is self-induced when the current in a circuit change.
- of motor: the electromotive force opposing motion which is induced in a conductor moving in a magnetic field.
- of polarization: a reduction of a cell's natural electromotive force which occurs when polarization produces a back electromotive force.

Background

Unwanted effects in physical measurements that must appear above a certain threshold in order to be measured. The phrase is frequently used to distinguish naturally occurring effects from those caused by human actions, such as the effect of *Cosmic Rays* in *Radiation* investigations.

Balance

A partial or complete state of equilibrium. A weighing scale or scales. There are many different kinds. A rigid beam is balanced on a *Fulcrum* at its halfway in the beam balance, and similar pans are suspended from each end of the beam. The weighing object is placed in one pan, while standard masses are added to the other until the beam is horizontal; the sum of the standard masses equals the unknown mass.

A rider is furnished with a sensitive beam balance, which is a small mass that can be moved along the top of the balance beam on which is B marked a calibrated scale. The beam of a sensitive balance sits on agate knife edges to reduce friction, and it is encased in a glass-fronted cabinet to lessen the effects of air currents and temperature fluctuations.

Ballistic Galvanometer

Electric *Charge* is measured with this instrument. The charge must travel through the galvanometer in a short amount of time relative to the *Period* of natural *Oscillation* of the galvanometer; the charge must pass before the galvanometer begins to deflect. The charge flowing is proportional to the initial deflection (throw) of a *Moving Coil Instrument.*

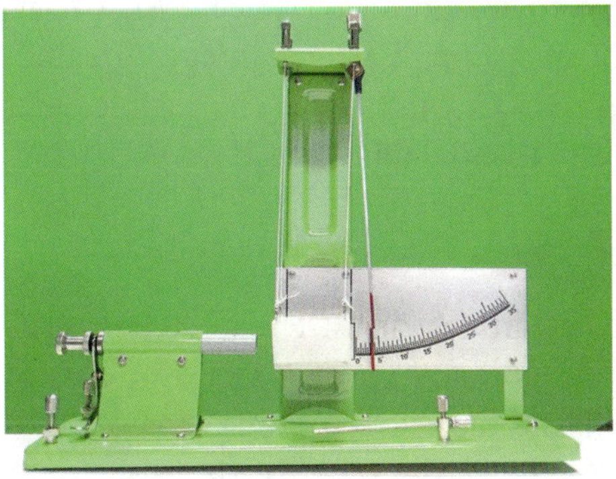

Figure 5. Ballistic Galvanometer.

Source: Image by Wikimedia commons.

Ballistic Pendulum

A pendulum used to measure the velocity of a projectile striking it The mass, m, of the freely suspended pendulum is very much greater than the mass, m', of the projectile. If u' is the velocity with which the projectile strikes the pendulum and assuming the projectile embeds in the pendulum, then by the CONSERVATION LAW for momentum

$$m'u' = (m + m')u,$$

where u is the velocity with which the pendulum begins to move. By the Conservation Law for energy

$$0.5(m + m')u^2 = g(m + m')h,$$

where h is the observed vertical height of rise of the centre of mass. Hence u' can be foun .

Ballistics

The study of projectile motion and propulsion. Internal ballistics is concerned with events that occur prior to launch, while exterior ballistics is concerned with events that occur after launch.

Balmer Series

A sequence in the hydrogen atom's SPECTRUM. $1/\lambda. = R(l/2^2 - l/n^2)$, where λ is the wavelength of a line, n is any integer greater than or equal to 3, and R is the RYDBERG CONSTANT. The visible section of the spectrum encompasses all of the lines in this series.

Band Spectrum

Band theory 21 occurs in a form of SPECTRUM that is characteristic of molecules. Each band is made up of a series of closely spaced lines. The band head is a highly defined edge on one side of a band. The intensity is zero on one side of the head, but steadily decreases on the other.

The band is said to be degraded towards the violet if the slow fall-off is on the shorter (longer) wavelength side (red). The wavelength of the band head is crucial for identifying the chemical. Different VIIJRATIONAL ENERGY LEVELS of the molecule give rise to the lines within a band; each band corresponds to a transition between a high and low electron ENERGY STATE.

Bandwidth

The band theory's key achievement is that it explains the variances in electric conductivity of different materials. Electron movement in the presence of an electric field causes conductivity; electron movement indicates that an electron is gaining energy from the field. According to QUANTUM THEORY, this gain can only happen in discrete amounts and equals the energy difference between two permissible states.

The banned band may not exist in metals and good conductors, i.e., the permissible bands overlap, allowing for easy energy gain. The forbidden band in electrical insulators is thought to be too wide for the electron to jump. It is suggested that the forbidden bind for an intrinsic semiconductor is thinner, allowing one electron to cross it.

The atoms of the impurity offer energy levels in the prohibited zone (band) in an IMPURITY SEMICONDUCTOR, which serve as stepping stones for the electrons.

Banking of Track

Track banking is a technique for increasing the safe cornering speed. The friction force at the points of contact with the track provides the necessary force towards the center of the circle when a vehicle moves around a circular path. With increasing speed, the needed force increases, and if friction is insufficient, the vehicle will move away from the center, causing it to skid.

The maximum safe speed can be raised by inclining the track to the horizontal (banking) such that the NORMAL response at each point of contact has a COMPONENT towards the center, which aids friction. This factor is proportional to the angle of banking's sine.

This is why racing tracks include saucer-shaped curves; a driver may then relocate to a section of the track that is sufficiently steep to prevent side slide while maintaining high speed. By leaning towards the center, a rider boosts his or her stability even more.

Barkhausen Effect

A sequence of discontinuous increments in the amount of FERROMAGNETISM produced by a continually changing magnetic field. The steps are understood as MAGNETIC DOMAINS alignment and can be shown by winding two coils on an iron core and gradually increasing the current through one coil. On a CATHODE RAY OSCILWSCOPE, the discontinuities in the INDUCED ELECTROMOTIVE FORCE in the other coil may be seen.

Barn

A unit of area equal to 10^{-28} square meter and is denoted by the symbol b. The cross sections of atomic nuclei are commonly expressed in bam in SCATTERING tests.

Barograph

An aneroid BAROMETER is a recording BAROMETER. In most cases, the pointer is holding a pen that is in contact with a moving chart. As a result, a continuous record of atmospheric pressure variation throughout time is obtained.

Barometer

A device that measures atmospheric pressure. There are various different kinds. Changes in atmospheric pressure cause movements in the thin corrugated top of an evacuated sealed metal box in the aneroid barometer. A train of levers transmits these movements to a pointer, and the pressure is read immediately on a graded scale that the pointer can move over. This is the sort of barometer found in most homes.

Figure 6. Barometer.

Source: Image by Wikimedia commons.

Barometer Conventions

The temperature and gravitational acceleration values at which a mercury BAROMETER can read directly in acceptable pressure units are 273 kelvin and 9.80665 meter per square second, respectively.

Barometric Reduction

The conversion of barometric measurements to their equivalent value at sea level. The goal is to draw meaningful comparisons between pressure values acquired from different altitude stations.

Barrier Potential Difference

A potential difference between two semiconductors at a p-n JUNCTION. It prevents electrons from flowing to the p-type semiconductor and positive HOLES from flowing to the n-type semiconductor. Its maximum voltage is approximately 0.1 volt.

Baryon

Any elementary particle made up of three quark. Baryons are unstable apart from the proton; their decay products invariably involve a proton.

B is the symbol for the baryon number. The value of a quantum number is I for a BARYON, -I for an antibaryon, and 0 for all other particles. In all particle interactions, the overall baryon number is conserved.

Base

The region between the EMITTER and the COLLECTOR of a TRANSISTOR. A geometric solid's bottom face. The bottom horizontal straight line of a geometric shape. In a positional number system, the number of units clustered together and denoted by I in the next place.

The decimal system, for example, has a base of ten, while the binary system has a base of two.

An electric CELL, or two or more such cells linked and used as a single unit, is referred to as a battery. A multi-cell example is the LEAD ACID BATTERY, which is widely used in automobiles. CELL WITH A SINGLE LECLANCHE Dry batteries is widely used in portable radios, flashlights, and other devices.

Beats

When two sound wave trains with the same amplitude but slightly different frequencies are stacked, this phenomena happens. INTERFERENCE occurs between them, resulting in a consequent vibration with amplitudes ranging from zero to twice that of either parent wave train. As a result, the ear registers a series of maxima, which are the beats. Due to the slight frequency difference between the two parent vibrations, two ULTRASONIC waves can produce audible beats.

Beckmann Thermometer

A mercury thermometer used to measure tiny temperature changes. The capillary tube has a scale of 5 K and is connected to a reservoir at the top of the thermometer. The scale has extremely fine graduations. The thermometer is used to measure temperature variations (rather than actual temperatures) in the range of 0 to 100°C. To do so, use the reservoir to adjust the amount of mercury in the bottom bulb to a level that is appropriate for the temperature level being studied.

Beta Decay

A form of RADIOACTIVE DECAY in which an unstable NUCLEUS ejects a POSITRON and a NEUTRINO, or an electron and an antineutrino. This modifies the ATOMIC NUMBER of the nucleus in question, but not its MASS NUMBER. RADIOACTIVITY is another term for radioactivity.

Betatron

A device that uses MAGNETIC INDUCTION to accelerate electrons. A torus-shaped evacuated chamber holds the electron beam, which is held between the pole components of a ring-shaped magnet. The magnetic field is designed to rise as the electrons' energy increases, keeping the electrons in a constant orbit. The magnitude of the magnetic field is proportional to the electron momentum. The top limit of energy attainable is around 500 million electro-volts, which is determined by electron radiation losses. Nuclear physics research, RADIOGRAPHY, and RADIOTHERAPY all use the equipment.

Beta Waves

In ELECTROENCEPHOLOGRAPHY, waves occur. They are usually spindle-shaped and have a frequency of 14–50 hertz, with a lesser amplitude than ALPHA WAVES. The beta waves entirely take over while dreaming. Studying the waves is most beneficial in diagnosis, for example epilepsy, tumors and accident damage.

Bifocals

Bifocals are spectacles that have two lenses. When the wearer is staring ahead at distant things, each upper portion is a DIVERGING LENS that corrects MYOPIA. When the wearer is gazing down at close work, each lower portion is a CONVERGING LENS that corrects PRESBYOPIA.

Bimetallic Strip

Roll welded together are two pieces of various metals with uneven COEFFICIENT OF EXPANSION, such as brass and iron. When the temperature changes, the strip bends due to unequal amounts of expansion. Temperature can be shown using these strips. They're also found in THERMOSTATS and safety devices that shut off an electric circuit when a certain temperature is reached.

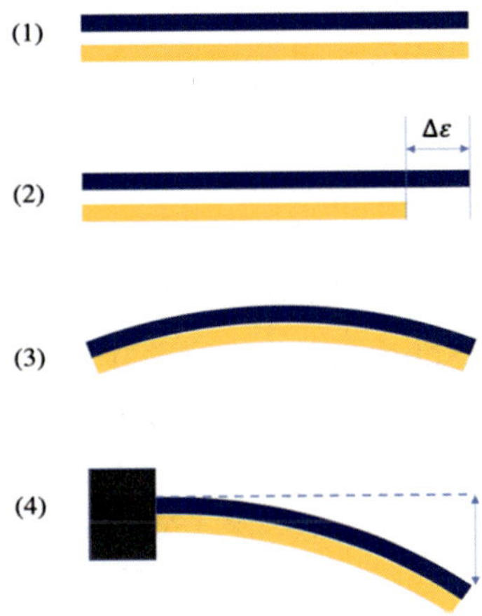

Figure 7. Bimetallic Strip

Source: Image by Wikimedia commons

Binary Notation

Only two digits, 0 and 1, are used to represent numbers. BINARY ARITHMETIC is another term for binary algebra.

Binary Star

Under the influence of their mutual gravitational attraction, a pair of stars revolves around their common CENTER OF MASS. One of the components of an astrometric binary star is too weak to be seen directly. Its presence is inferred by changes in the visible component's motion.

The components of a near binary star are too close together to be viewed independently. Their orbits could differ from the standard two-body orbits. As seen from Earth, an eclipsing binary star is made up of two components, each of which passes in front of the other for one orbital period. The system's brightness varies as a result of this.

Binding Power

The amount of energy needed to separate a particle from its parent system, such as an electron from an atom. The amount of energy needed to break down a system into its constituent particles, such as a nucleus into NUCLEONS.

Binoculars

A pair of terrestrial TELESCOPES designed to be used with both eyes at the same time, with focusing tubes controlled by a single screw adjustment. Prism binoculars have a prism system in each telescope, which produces an upright image; additionally, the tube length is shorter than an ordinary telescope of equal power.

Binocular vision is the ability to see with both eyes open.

The use of both eyes at the same time. Due to the separation of the eyes, two images from somewhat different perspectives are obtained. The brain of the observer combines the two images to create the illusion of depth.

Bioluminescence

LUMINESCENCE is a property of living organisms such as bacteria, fireflies, glow worms, and a variety of deep-sea species.

Biophysics

The application of physics' tools and concepts to the study of biological systems. It could be about the physics of the systems themselves, the biological consequences of physical agents, or the application of physical methodologies to the study of biological problems.

Bit

The bare minimum of information required to define one of two possible states. In BINARY ARITHMETIC, for example, each of the two digits 0 and 1 represents a bit.

Black Body

A body that absorbs all of the radiation it receives and does not reflect any of it. At any temperature, the amount of incident radiation equals the amount of emitted radiation to achieve equilibrium. The PLANCK RADIATION FORMULA describes the spectral distribution of emitted radiation. In actuality, no substance is an ideal black body; the radiation emitted from a small hole or slit in a constant temperature enclosure is the closest realistic approximation to black body radiation.

Black Body Temperature

A radiation PYROMETER is used to determine a body's temperature. It is usually significantly lower than the body's real temperature.

Black Hole

The hypothetical state of a star that has exploded, i.e., turned into a SUPERNOVA, and left a core with a mass greater than three times that of the Sun. The consequence of gravitational forces on such a core is to cause it to compress heavily, resulting in a gravitational field that is big enough to prevent any matter or radiation from escaping.

Work on the BINARY STAR Cygnus Xl, which is a strong producer of X RAYS, has provided some observational evidence for the black hole theory; this system consists of a SUPERGIANT and a little invisible partner (black hole?) of mass around ten times that of the Sun. The X-ray radiation could be explained by the supposed black hole constantly attracting stuff from the supergiant.

Blink Comparator

A viewing device that allows the user to alternate between two images of the same section of the sky taken at separate times. A pulse in brightness shows the presence of a variable star, whereas a change in position of an image appears as an apparent movement. As a result, the device is ideal for detecting location and brightness variations without the need for precise measurements.

Blood Flow Measurement

Nowadays, the potential difference induced across the artery walls by blood flow is measured using an electromagnetic measurement in which a magnetic field of square or sawtooth wave form is applied to a bared artery and the potential difference induced across the artery walls by blood flow is recorded. The induced electromotive force in the leads is in the form of the rate of change of the magnetic field, whereas the induced electromotive force owing to blood flow is in the form of the magnetic field.

Because the flow and lead electromotive forces have different wave patterns with a square or sawtooth magnetic field wave form than a sinusoidal magnetic field, the flow electromotive force may be easily separated. Unlike prior methods of blood flow measurement, the electromagnetic method does not necessitate an artery incision.

Blood Pressure

The force exerted by blood on the artery walls. It is determined by a number of physical factors, including the flexibility of the artery walls, capillary resistance, and blood volume and viscosity. A SPHYGMOMANOMETER is used to measure the pressure.

Boiling Point

At what temperature does a liquid's SATURATED VAPOUR PRESSURE equal the external pressure? The boiling point is determined by intermolecular forces; the stronger these forces are, the higher the boiling point. With an increase in external pressure, the boiling point rises as well. The normal boiling point is the value of the boiling point when the external pressure is one standard atmosphere.

Boiling Water Reactor

A NUCLEAR REACTOR that uses water as both a moderator and a coolant. TYPES OF NUCLEAR REACTORS is another term for nuclear reactors.

Boltzmann Constant

The UNIVERSAL GAS CONSTANT per atom is denoted by the symbol k. (or per molecule in the case of a molecular gas). It is thus the AVOGADRO CONSTANT divided by the universal gas constant per MOLE. Its joule per kelvin value is 1.38054×10^{-23}.

Bomb Calorimeter

A calorimeter is a device that measures the HEAT OF COMBUSTION. A known mass specimen is placed in a tiny capsule that can be sealed in a steel vessel. The sealed jar is then placed in a lagged vessel with a known mass of water at a certain temperature, and oxygen is supplied at roughly 20 atmospheres pressure. The water is heated to its maximum temperature after the specimen is ignited electrically.

Bond

The inclination of atoms to prefer electronic configurations in which at least the outer electron shell is complete leads to a union between them. coordinate bond; covalent bond; electrovalent bond; metallic bond are other terms for the same thing.

Born Approximation

The interactions are treated as perturbations of free particle systems in this method of estimating cross sections in scattering issues.

Bose-Einstein Condensation

The condensation of a vapor whose molecules follow BOSE-EINSTEIN STATISTICS to a condition where some of the molecules have a momentum of almost zero rather than a wide range of values.

Bose-Einstein Statistics

When any number of particles can occupy each energy state, Quantum Statistics is used to the distribution of comparable particles between multiple permissible energy values. When the temperature is high enough, i.e., when a significant number of energy levels are stimulated, the statistics lead to the classical situation. Distribution Law Maxwell-Boltzmann.

Boson

A particle that follows the Bos-Einstein Statistics but not the Pauli Exclusion Principle. Alpha Particles, Photons, Pions, and all nuclei of even mass number are examples.

Boundary Layer

The layer of incompressible fluid that forms next to a stationary object and flows past it, with a thickness of 50 micrometers to 100 riometers. For the study of lift and drag, the behavior of this layer is crucial.

Brackett Series

A sequence in the hydrogen atom's SPECTRUM. The formula is $1/\lambda = R(1/4^2 - 1/n^2)$, where λ is the line's wavelength, n is any integer higher than or equal to 5, and R is the RYDBERG CONSTANT. The lines are visible in the infrared spectrum.

Breakdown Voltage

The voltage at which an insulator can no longer resist the voltage and conducts. The word is frequently used to describe the flow of electric current through a gas.

Bridge

A four-terminal electrical network can be any of a number of different types. A detector is connected across the output terminals and an e.m.f. is applied to the input terminals. The detector reading can be adjusted to zero by altering the values of one or more network components.

The values of the components are then related in certain ways, allowing one component to be assessed in terms of the others. The WHEATSTONE BRIDGE, as well as the CAREY-FOSTER METHOD, are used to measure resistance; the KELVIN DOUBLE BRIDGE, with eight terminals, is used to compare low-resistance standards.

Capacitance is measured using the CAPACITANCE BRIDGE. The ANDERSON BRIDGE is used to compare self-inductance and capacitance. The OWEN'S BRIDGE or Maxwell's BRIDGE are acceptable; the CAMPBELL BRIDGE can be used to compare mutual inductance with capacitance.

Brinell Test

A test that involves driving a tiny hard steel ball into the surface of a metal to determine its hardness. The Brinell number is the ratio of the load on the ball to the area of the depression generated, which is used to express hardness.

Brush Discharge

In a gas, an electric discharge near high-voltage sharp points. The voltage required is lower than that needed for a SPARK DISCHARGE or an ARC. At the point's tip, the discharge appears as brief bright streamers.

Figure 8. Brush Chamber.

Source: Image by Wikimedia commons.

Bubble Chamber

The tracks of ionizing radiation can be seen using this instrument. It comprises of a vessel filled with a liquid (usually hydrogen or helium) under pressure at a temperature above the boiling point of the liquid. Superheating is accomplished by reducing the pressure on the liquid for a few moments while taking a photograph. Before the liquid boils, the pressure is reapplied. During superheating, particles traveling through the liquid leave a trail of tiny bubbles.

Bumping

When a liquid is heated to a temperature higher than its natural boiling point, it boils violently. As a result, the bubbles have a higher pressure than the surrounding air.

Bunsen Grease Spot Photometer

A tool for comparing the luminous intensity ratings of two light sources. The grease spot is created on an opaque white screen and held in place by the two sources. The energy received at the screen from each source is the same when juxtaposed mirror pictures of the two sides of the spot appear identical. As a result, the source intensities can be compared using the Inverse Square Law.

C

Cadmium Sulfide Cell

A PHOTOCELL is made up of a layer of cadmium sulfide sandwiched between two electrodes. The cell is utilized in exposure meters and is substantially more sensitive than a SELENIUM CELL.

Calculus

The DIFFERENTIATION and INTEGRATION of functions is a branch of mathematics.

Calculus of Variations

The study of definite integrals' maximum and minimum qualities. The method can be used to solve physical problems such as determining equilibrium conditions using the LEAST ENERGY PRINCIPLE, determining the course of a light ray using F'ERMATS PRINCIPLE OF LEAST TIME, and solving dynamical problems using the HAMILTON PRINCIPLE.

Calibration is the process of taking readings from an instrument for standard or known quantities in order to convert arbitrary readings from the device to absolute values.

Callier Coefficient

The ratio of a photographic negative's density measured with parallel light to its density recorded with completely diffuse light. The average is approximately 1.5.

Calomel Electrode

Calomel is a mercury electrode in contact with a potassium chloride solution saturated with mercurous chloride. As a reference electrode, it is employed.

Caloric Theory

The concept of heat as a weightless fluid is the basis for this hypothesis. It was supplanted in the mid-nineteenth century when heat was demonstrated to be a type of energy, as it fails to account for the production of an infinite amount of heat by friction.

Calorie

At standard pressure, the amount of heat necessary to raise the temperature of one gram of water from 14.5°C to 15.5°C. In scientific computations, the unit is now obsolete.

Calorific Value

The quantity of heat released when a substance's unit mass is completely burned. It is anticipated that any steam created will condense. The measurement can be made in a BOMB CALORIMETER

Calorimeter

A device for taking quantitative temperature measurements. The METHOD OF MIXTURES uses the simplest form, which consists of a copper can with a lid through which a thermometer and stirrer pass; the can is wrapped in insulating material like cotton wool and placed within a bigger can.

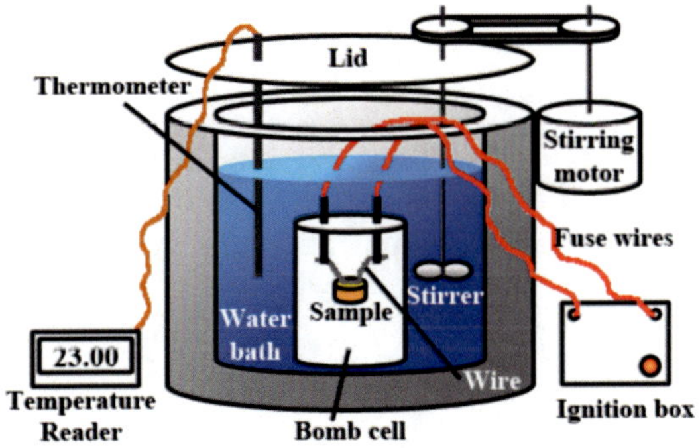

Figure 9. Calorimeter.

Source: Image by Wikimedia commons.

Calorimetry

Specific heat capacity, calorific value of a fuel, heat of atomization, or heat of combustion are all examples of heat measurement.

Camera

A device that creates a photographic image. It's just a lighttight box with a shuttered convex lens on one side that focuses a genuine picture on a photographic film or plate behind the lens when the shutter is released. The camera is ready for the next exposure after winding on the film or changing the plate.

Camera Obscura

A convex lens or a tiny aperture is used to create an image of an exterior scene within a box or chamber. In the early SUNSPOT research, such a gadget was employed.

Campbell Bridge

In terms of mutual inductance, an electrical BRIDGE for measuring capacitance (C). A microphone or an oscilloscope could be used as the detector. The values of the variable resistors R, R_1, and R_2 are modified until the detector output is zero. $L/M = (R + R_1)/R$ and $M/C=RR_2$, where L is the self-inductance.

Canonical Equations

In classical mechanics, equations represent the motion of a group of particles. $dq_n/dt = dH/d_{pn}$ and $dp_n/dt = -dH/d_{qn}$, where p_n and q_n are the nth generalized momentum and position co-ordinates at time t, respectively, and H is the HAMILTONIAN.

Capacitance

The ratio of an object's electric charge to the change in its potential as a result of that charge. The farad is a unit of measurement that is equivalent to one coulomb per volt. The capacitance of a given body is constant; the sphere capacitance is $4\pi\varepsilon_0 a$ for an isolated conducting sphere of radius a, where ε_0 is the absolute permittivity for a vacuum.

Capacitance Bridge

A special sort of WHEATSTONE BRIDGE that can be used to measure capacitance. C1 is a CAPACITOR with an unknown capacitance, while the capacitance values of capacitors C2, C3, and C4 are known, with the exception of c2, which is adjustable. When the key, K, is depressed, the capacitance of C2 is changed until the quadrant electrometer, E, exhibits no deflection. The capacitance of C1 can then be calculated by multiplying

C/C2 by CJC4. (A capacitor is usually denoted by the same symbol as its capacitance)

Capacitor

It consists of one or more pairs of insulator-separated conductors and is used to store electric charge. The electrodes or plates are the conductors, and the dielectric is the insulator, which can be solid, liquid, or gaseous.

Capacitor Microphone

A MICROPHONE in which one plate of a CAPACITOR is formed by a diaphragm, while the other plate is fixed. Capacitance changes are caused by diaphragm movements caused by soundwave pressure differences. Changes in capacitance cause comparable changes in potential difference across the capacitor since the microphone is connected in series with a high resistance across a stable potential difference of roughly 300 volts. The microphone has a decent frequency response and is free of background noise, but it suffers from directional effects and poor output.

Capture

Any process in which an atom, ion, molecule, or atomic nucleus acquires another particle. In radiative capture, for example, a nucleus captures a neutron and then generates a gamma ray; in K capture, a nucleus acquires a K shell electron and then emits a neutrino, maybe followed by an electron from an outer shell filling the K shell vacancy and emitting an X ray photon.

Carat

- 2×10^{-4} kilogram, one of the TROY UNITS OF MASS.
- In a 24-carat gold alloy, the number of parts by mass of gold. Thus, 24 carat gold is pure gold; 20 carat gold contains 20/24 pure gold and 4/24 additional metals.

Carbon Cycle

The production of one helium nucleus from four hydrogen nuclei as a result of a series of six nuclear events, with carbon acting as a catalyst.

Carbon Dating

For organic remnants, a radioactive dating method is used. The majority of carbon in living plants and organisms is 12C, but a minor quantity of the radioactive 14C isotope, with a half-life of 5760 years, is also present. Because there is no renewal of radioactive carbon in dead tissue, the isotope ratio and thus the radioactivity displayed decreases exponentially over time.

Carbon Microphone

Figure 10. Carbon Microphone.

Source: Image by Wikimedia commons.

A microphone that works by varying the electric resistance of carbon in response to pressure changes in sound waves impacting on the microphone.

Carrier

(1) An electron or HOLE traveling through a metal or semiconductor, allowing charge to pass through the substance and so causing conductivity.

(2) The non-radioactive version of a chemical substance that is used to introduce the radioactive form into a system for TRACER research.

Carrier Wave

In radio transmission, a radio wave with a constant amplitude and frequency that is MODULATED by the signal.

Cascade

A cascade of electrical circuits is a procedure or piece of equipment that is set up in a number of steps with the output of one stage becoming the input of the next.

Cascade Liquefaction

A liquefaction method in which a gas with a high critical temperature is liquefied by increasing pressure, then allowed to evaporate under lower pressure, lowering the temperature. This cooling effect is employed to lower a second gas's temperature below its critical temperature, which is lower than the first gas's. The compression evaporation process cools the second gas in the tum, which then cools the third gas, and so on. This technique can yield liquid oxygen.

Cascade Shower

The continuous making of electrons and high-energy photons as a consequence of high-energy particles coming into contact with dense matter, i.e., via bremsstrahlung and pair production.

Cassegrainian Telescope

An astronomical telescopic reflector built through combining a primary concave mirror and secondary convex mirror. The light hits on the large concave mirror with a hole at the center, through which the light is viewed after being reflected, initially at the paraboloid concave mirror and later at the hyperboloid convex mirror.

Cathetometer

An instrument for gauging vertical distances. It is made up of an accurately graduated scale and a telescope placed horizontally in order to move it up and down the fixed vertical scale.

Cathode Follower

An electronically amplified radio circuit in which the input is between the grid and the remote end of the cathode circuit and the output is taken between the cathode and the ground. The cathode follower is not a voltage amplifier, yet, it amplifies current over a wide frequency range. The input and output voltages are in phase; the input impedance is high and the output impedance is low.

Cathode Ray Oscilloscope

A common laboratory instrument whereby electrical signals can be visually displayed on a CATHODE RAY TUBE. The signal under investigation is applied, after amplification, to the vertically deflecting plates of the tube; the amount of deflection of the spot is proportional to the signal strength.

The spot is also moved horizontally across the screen by a TIME BASE generator incorporated in the oscilloscope. The resultant trace observed on the screen is therefore a composite of the two voltages and is thus a graph of signal against time. The commonest type of time base is of adjustable frequency and has a SAWTOOTH WAVE FORM so that after traversing the screen at constant speed the spot flies back almost instantaneously to its starting point.

Causality

The relationship between causes and effects. The Heisenberg uncertainty principle states that it is not possible to measure and identify a variable of a subatomic particle, whereas in quantum physics, the concept of causality states that there are chances or probabilities that certain subatomic particles may exist in specific positions and participate in specific events.

Caustic

The curve or surface to which nonaxial light rays, after being either reflected or refracted from a curved surface, become tangent. The cusp, i.e., the apex of the caustic, is the focal point for paraxial rays.

Cavitation

The phenomenon that involves the formation of tiny vapor bubbles in a liquid at areas of low-pressure that form in regions with high-velocities, for instance, inside centrifugal pumps, water turbines, and marine propellers.

As a result, surface pitting occurs, which automatically subsides when the bubbles enter an area of higher pressure.

Cavity Resonator

An electronic device with a space that is confined by metallic walls inside which an electromagnetic field is formed for usage. They are an important component of microwaves and optical systems. The device has numerous resonant radio frequencies and their values are determined by the cavity dimensions. The main uses are at frequencies greater than 300 megahertz.

Celestial Mechanics

The branch of astronomy devoted to the study of the relative motion of objects in outer space due to the forces of gravity. It was founded by Sir Isaac Newton.

Cell

An individual electrical energy device that produces electric energy by transforming and converting chemical energy directly into electric energy. It consists of two electrodes dipping into an electrolyte; ions are produced or discharged at each electrode so that one electrode becomes positively charged and the other negatively charged.

Once the connection is established via a circuit, the flow of current occurs and once a cell is fully assembled after passing current, it is called a primary cell. A secondary cell, on the other hand, needs to be charged after being assembled, i.e., a current has to pass through it in the opposite direction to its discharge; the chemical actions in this cell are reversible.

Centigrade Scale

Also known as Celsius temperature scale is the scale in which the melting point of ice is 0° and the boiling point of water 100°. The formula used for converting a temperature from its representation on the Fahrenheit (F) scale to the Celsius (C) value is $C = 5/9(F - 32)$.

Centered Optical System

An optical system in which the vertex centers of curvatures are formed on a straight line. The number of spherical reflecting or refracting surfaces has their focal points on a common axis. In order to measure the position, nature and size of an image formed by rays close to the axis in any such system, the positions of only six special axial points; known as cardinal points are to be identified.

Center of Buoyancy

The point at which the buoyancy force is applied on the body and is the geometric center of the submerged part of the body.

Center of Curvature

The center of a circle that passes through a curvature at a specific point and forms the same tangent and curvature at that point and its radius is equal to the radius of curvature of the curve at that point.

Center of Gravity

The imaginary or theoretical point at which the total weight of the body is thought to be concentrated due to the gravitational field. In cases where the gravitational field is found to be constant, the gravitational forces acting upon each particle become parallel and as a result the force passes through a point fixed relative to the body, i.e., through the 'center of mass,' and the center of gravity corresponds to the exact same point.

In cases, where the gravitational field is non-uniform, the resultant forces on a body are typically equal to a single force and a couple; the single force does not generally pass through a point fixed relative to the body for all positions of the body and so therefore the center of gravity ceases to exist.

Center of Mass

Also called the balance point, it is the point at which the weighted relative position of a distributed comes to zero. Therefore, all the particles that form the body of a product of particle mass and particle perpendicular distance from any plane through the point is zero.

Centrifugation Separation

A technique employed for separating particles from a solution as per their size, shape, and density, viscosity of the medium and rotor speed. A centrifuge tube is used to separate the particles.

Centrifuge

A laboratory device that is a bar or flywheel mounted atop a vertical axle at which it can rotate rapidly. It is used for the separation of fluids, gas or liquids on the basis of their density. The separation process takes place in tubes that contain the specimens to be separated and are attached to the extremities of the wheel or bar in a manner that they are free to pivot outwards at high speed.

The forces produced greatly exceed the gravitational force and so the settling of suspensions can be speeded up. A common example of centrifugation is the extraction of fat from milk in order to produce skimmed milk, the removal of water from moist lettuce with the help of a salad spinner, or the spin-drying of water in washing machines for removing water from the clothes.

Centripetal Force

A net force that makes a body follows a curved or circular path. The force is aimed at the center of the circle and is of magnitude mv^2/r, where v is the speed of the particle. The body applies a centrifugal force, equivalent in capacity but in the opposite direction to the centripetal force, on the object that exerts the centripetal force.

Centroid

A principle of mathematics and physics is a geometric center of a triangle where the three medians coincide. Therefore, it corresponds to the center of mass of a homogeneous solid body of shape identical to the geometric one. The centroid theorem states that the centroid of the triangle is at 2/3 of the distance from the vertex to the mid-point of the sides.

Cherenkov Radiation

An electromagnetic radiation produced when a charged particle (such as an electron) passes through a dielectric medium at a speed greater than the phase velocity (speed of propagation of a wave in a medium) of light in that medium. The bluish light formed thereafter, is emitted by a beam of high-energy charged particles when travelling through a medium of refractive index n at a speed v which is greater than the speed of light in the medium.

It is emitted in a conical wave front whose axis lies in the beam direction. The cone's apex angle is 29 where cos 9 = c/(nv) c is the speed of light in empty space; v may thus be found by measuring 9. The Cerenkov radiation is an electromagnetic shock wave, i.e., it is the optical analog of the sonic boom.

CGS Units

The centimeter–gram–second system of units (abbreviated CGS or cgs) is a variant of the metric system based on the centimeter as the unit of length, the gram as the unit of mass, and the second as the unit of time. All CGS mechanical units are unambiguously derived from these three base units, but there are several different ways in which the CGS system was extended to cover electromagnetism.

The system was extended to cover thermal measurements by the addition of the inconsistently defined calorie. Two methods of further extending the system to embrace electrical measurements were developed; they involved the introduction of electromagnetic units and electrostatic units. All versions of the CGS system have now been superseded by SI units.

Chain Reaction

A sequence of reactions in which a reactive product or by-product leads to additional reactions to occur. It is a process in which neutrons released in fission produce an additional fission in at least one further nucleus.

This nucleus in turn produces neutrons, and the process continues repeatedly. The process may be controlled (nuclear power) or uncontrolled (nuclear weapons). For example, the fission of a uranium–235 nucleus is accompanied by the emission of from one to three neutrons, any of which is capable of causing fission in another 235U nucleus.

Change of State

The change in the state of a substance from on state to another state. Examples include changes brought about by the process of boiling, condensing, freezing, melting or subliming.

Channel

A fixed band of frequencies, or a specific path, engaged in the transmission and reception of electric signals. The region in a field effect transistor between source and drain whose conductivity is regulated by the amount of voltage applied at the gate.

Chaotic System

A system in which, for any sort of motion, minor differences in initial conditions can result in hugely different results. An example is weather experienced in England.

Characteristic Curve

A graph that highlights the relationship between two variables for a device. An example is the plot of photographic density against logarithm of exposure for specified conditions of processing a particular photographic emulsion; another is the plots of voltage against current for various electrical and electronic devices.

Characteristic Function

It is one out of a set of functions satisfying a particular equation with specified boundary conditions. In wave mechanics, characteristic functions are physically possible solutions of Schrodinger's equation for an atomic particle. Characteristic functions are also a frequent feature in quantum mechanics known as Eigen function.

Characteristic Value

The value of a specific parameter associated with a characteristic function. In cases wherein more than one solution is present for a differential equation corresponding to a particular characteristic value, the system is said to be degenerate. If the characteristic value is associated with an Eigen function, it is referred to as an eigenvalue.

Characteristic X Radiation.

A series of discrete-wavelength X-rays characteristic of the emitting element and independent of its state of chemical combination.

Charge

In physics, charge, also referred to as electric charge, electrical charge, or electrostatic charge and symbolized q, is a property of a unit of matter that indicates the extent to which it has more or fewer electrons than protons, If there are more electrons than protons, the atom has a negative charge. Charge is conventionally designated positive or negative such that like charges repel one another and unlike charges attract. The force is thought to result from the exchange of photons between charged particles.

The natural unit of negative charge is that possessed by the electron; the proton carries a positive charge of equal magnitude. The charge on the electron, symbol e, is determined by applying Millikan's electronic charge determination method.

Charge Conjugation Parity

It is a quantum number attributed to the elementary particles of zero charge, baryon number, and strangeness. It is conserved in both strong and electromagnetic interaction.

Charged Particle

It refers to any particular particle that carries a positive or negative charge. Electrons and protons are common examples of charged particles. Once a charged particle enters a magnetic field whose direction is perpendicular to the direction of motion of the particle, the particle is deflected as indicated its path becoming circular while in the field.

It will be seen that Fleming's left-hand rule is obeyed, bearing in mind that the conventional current direction is opposite to the direction of motion of negatively charged particles and in the same direction as the motion of positively charged particles. If the particle speed is v, its charge q, and the magnitude of the magnetic induction B, then the magnitude of the force on the particle is Bqv and its direction is perpendicular to the particle path and to the magnetic field.

Charge To Mass Ratio of Electron

A quantity first measured by Thomson using the apparatus. A beam of electrons was subjected to magnetic and electric fields such that they deflected the beam in opposite directions; the strengths of the deflecting fields were adjusted to give zero deflection of the spot where the beam struck the fluorescent screen. It can then be shown that $e/m = ED / (B^2Ll)$.

Charge Transfer Device

A semiconductor device in which discrete packets of charge are transferred from one location to the next; the device can be used for short-term storage of charge in a particular location.

Charging By Induction

The process in which a conductor is charged by a charged body in the absence of any physical contact being established between them. The charged body attracts charge of opposite sign to its own on the closest parts of an uncharged conductor to which it is near; charges of the same sign are repelled to the remotest parts of the conductor.

Momentary earthing removes the latter charge, and on withdrawing the charged body the remaining charge on the conductor spreads over its surface. The conductor has thus been given a permanent charge of opposite sign to that of the original charged body and without loss of charge of the latter.

Chemical Energy

It refers to the energy conserved in the chemical bonds in any substance and which is further converted into a new form after a chemical reaction takes place. For example, chemical energy is converted into heat when a substance is burnt.

Chemical Hygrometer

An instrument used to measure the amount of water vapor in air, in soil, or in confined spaces. It contains drying tubes of known mass through which a known volume of air is drawn; the increase in mass of the tubes gives the amount of water vapor present. The experiment is repeated again for an equal volume of air which has been passed through water before entering the drying tubes. The ratio of the first mass to the second is established as the relative humidity of the atmosphere.

Chip

A very small piece of semiconductor consisting of a component, for example a resistor, transistor etc., or an integrated circuit.

Chi-Squared Test

A statistical significance test applied by summing the square of the difference of a result from the expected value, divided by the expected value, for all the results. The sum is denoted by x2; the smaller it is, the greater the agreement between observed and expected values.

Choke

It is an inductor that allows impedance to the flow of alternating current in a circuit. It is generally used to smooth the output of a rectifying circuit. A groove cut into the metal surface of a wave guide to a quarter-wavelength depth. It prevents the escape of microwave energy.

Chromatic Aberration

An aberration of lenses occurring due to the dependence of the refractive index of a material on wavelength. The image of a point source of white light therefore has a blue surround at the focus for red light and a red surround at the focus for blue light. At a position between the blue and red foci the image is white circle, known as the circle of least confusion.

Chromaticity Chart

A plane diagram obtained by plotting one of the three chromaticity co-ordinates against either of the other two. The locus of all visible wavelengths is horseshoe shaped. The straight line joining its ends is the locus of purples.

A point representing any other color lies within the figure, the point W representing white. The wavelength at which the line from W passing through a point representing any other color cuts the curve is the dominant wavelength for that color. The ratio of the distance of the color point from W to the distance of the dominant wavelength point from W is known as the excitation purity for the color.

Chromatography

A laboratory process of separating the components of a particular mixture. The different components of the mixture travel through the stationary phase at different speeds, thereby separating from each other.

Chromosphere

The inner gaseous atmosphere of the sun, extending about 15,000 meter beyond the photosphere.

Chronon

The time taken by a photon to travel a distance equal to the diameter of the electron; the time is about 10^{-24} second.

Circle

The locus of a point moving so that its distance from another point remains constant. If r is the radius of the circle then its circumference is 2nr and its area nil. An ellipse of zero eccentricity is a circle.

Circuit

An arrangement of electrical conductors, for example resistors or inductors, connected together to form a conducting path. When current passes a closed circuit is formed; in the absence of current an open circuit exists.

Circular Motion

The movement of an object along the circumference of a circle or rotating along a circular path. If a body moves With constant speed v round a circle of radius r, it has an acceleration of magnitude v^2/r directed towards the center of the circle. Since v = rw where w is the body's angular speed, the acceleration magnitude may also be written as rw^2.

Circumcenter

The center of a triangles' circumcircle. It can be found as the intersection of the perpendicular bisectors.

Circumpolar

Denoting stars or constellations which are always above the horizon when observed from a given geographical latitude.

Cladding

The process of bonding one metal to another to prevent corrosion. It involves the application of one material over another to form a surface or a skin or layer. The material used to enclose the fuel in a nuclear reactor. It is designed to prevent the escape of fission products and to support the fuel.

Class A Amplifier

An amplifier operating so that output current flows during the whole of the input cycle. It has low distortion and low efficiency.

Class AB Amplifier

An amplifier operating so that output current flows for between one half and the whole of the input cycle. It tends to operate as class A amplifier for low input signal levels and as a class B amplifier for high ones.

Class B Amplifier

An amplifier operating so that output current flows for only half the input cycle. It is efficient but produces some distortion.

Class C Amplifier

An amplifier operating so that output current flows for less than half the input cycle. It is efficient but produces appreciable distortion.

Class D Amplifier

An amplifier operating through pulse modulation. The input signal modulates the mark space. The ratio of a square wave causing it to operate a switching system. The resulting output current is proportional to the mark space ratio and therefore to the input signal. Such amplifiers are very efficient but switches operating sufficiently fast to avoid distortion are generally impracticable.

Clinical Thermometer

Clinical thermometer is used to measure the temperature of the human body. Mercury in glass thermometer reflects the body temperature. Unlike other types of mercury in glass thermometer the clinical thermometer has a constriction in the capillary just above the bulb.

This leads to ease of reading, the thread of mercury remaining in position after reaching its maximum value. The mercury may be returned to the bulb by shaking the thermometer. The calibration range is usually 35°–43°C, i.e., 95°–110° F; normal human body temperature is 36.9°C, i.e., 98.4° F.

Clocks

It is a device used for measuring time, generally using a periodic process of constant frequency. The more important types are as follows. The pendulum clock makes use of the fact that the period of a pendulum is a function of its length only; such a clock is accurate to about 0.01 second per day. The quartz clock has a regulating device in the form of an oscillator controlled by a crystal of quartz, which is kept in resonant vibration by means of the piezoelectric effect; the clock is accurate to about 0.001 second per day.

Closed Circuit Television

A television system in which camera, control system and receiver are cable linked so that there are no aerials or open circuits. It is used to observe a process from a central control room. A common example is the X-ray baggage inspection devices at airports.

Close Packed Structure

A structure obtained when atoms, assumed to be spheres of equal radius, are packed together so as to occupy a minimum volume. The structure is either face centered cubic or close packed hexagonal or combinations of these. Each atom is symmetrically surrounded by 12 others.

Cloud

A mass of water droplets or ice crystals suspended in the atmosphere and appearing as an opaque drifting body. Clouds form as a result of condensation of water in the sky. Minute atmospheric particles such as smoke encourage the formation of drops. Clouds are usually classified by their appearance and height.

Cloud Chamber

An instrument for making the tracks of ionizing particles visible as rows of little droplets, formed by condensation from a supersaturated vapor. The supersaturation may be produced for a short time by rapid expansion of the vapor causing adiabatic cooling. Continuous supersaturation is obtained in the diffusion cloud chamber.

Clausius Column

An apparatus used for separating gaseous isotopes. It consists of a vertical column many meters high with a heated wire running down its axis. A radial temperature gradient builds up between the wire and the tube wall. As a consequence of thermal diffusion, heavier molecules gather around the wire and lighter ones scatter towards the wall; the lighter isotope collects at the top of the tube.

Cluster

A collection of stars moving in the same direction. There are two main types. An open cluster comprises relatively small numbers of widely spaced young stars and is found in the spiral arms of the Galaxy. A globular cluster consists of enormous numbers of closely packed older stars and is found in the spherical halo surrounding the central nucleus of the Galaxy

Coaxial Cable

A cable consisting of a wire enclosed in an insulating coating several millimeter thick, which in turn is surrounded by a coaxial conducting cylinder which is often earthed. Coaxial cables do not produce external fields and are not affected by them and so are used for the transmission of high-frequency signals.

Figure 11. Coaxial Cable.

Source: Image by Pixabay.

Coefficient of Contraction

The ratio of the area of the vena contract A of a fluid jet to the area of the orifice through which it is discharging; values lie between 0.5 and 1.

Coefficient of Expansion

The fractional increase in size of a substance per unit temperature rise; the unit is thus K^{-1} or $^{o}C^{-1}$. A solid has three coefficients of expansion: α, which refers to length and is known as the coefficient of linear expansion; β. which refers to area and is known as the coefficient of superficial expansion; γ, which refers to volume and is known as the coefficient of cubic expansion or the volume expansivity.

Coefficient of Restitution

Symbol e. The ratio of the component of the relative velocity before impact, along the line of centers of two spheres, to the component of the relative velocity along the same line, in the opposite direction, after impact. The ratio is constant and depends only on the material of the spheres; for perfectly elastic material the ratio is 1; for absolutely inelastic material it is 0. For other material it lies between 0 and 1; for cxamplc, for ivory it is very nearly 1 and for putty it is almost 0. The value of e varies slightly at high relative velocity.

Coercive Force

The magnetic field strength required to remove the residual magnetism of a ferromagnetic material. If the substance is initially saturated the coercive force is known as the coercivity.

Coherent Radiation

It refers to the electromagnetic radiation of an individual phase. Most practical radiation sources are not coherent over a significant period of time since each emits large numbers of waves all out of phase with each other. Two sources of electromagnetic radiation are said to be coherent if they produce in-phase wave forms.

Coherent Units

A system of units in which a derived unit is produced by multiplying or dividing base units without introduction of a numerical factor; the base units are an arbitrarily defined set of physical quantities.SI units are coherent.

Cohesion

A type of force between molecules. For example, when a liquid is split, the greater the force of cohesion between the molecules the greater the chance of drop formation (rather than spreading over the receiving surface).

Coincidence Circuit

A circuit with two or more input terminals, all of which must receive a pulse within a specified time interval if an output pulse is to occur.

D

Dam

A structure designed to control the flow of water. It can have several uses: raising a water level to assist navigation, diverting water, generating hydroelectric power, storing water, land irrigation, flood control, providing recreation. Concrete dams are usually built with a convex face in contact with the water, thus making use of the arch principle of strengthening. Gravity dams are usually massive structures of earth and rock, presenting a flat face to the water and having a greater thickness at the bottom than at the top.

Damping

A decrease in the amplitude of oscillations with time. It is produced by forces resisting the motion, for example friction and electromagnetic forces. If after disturbance there is a large number of oscillations before the undisturbed state is regained, under damping is said to occur.

In the case of critical damping; the time taken to regain the undisturbed state is a minimum since oscillation just fails to occur. For over damping there is likewise no oscillation, but a longer time is required to regain the undisturbed state than for critical damping. In many measuring instruments, critical damping is often intentionally introduced in order to facilitate the observation of readings.

Daniell Cell

A primary CELL in which the positive pole is a copper rod which is immersed in copper sulphate solution and the negative pole is an amalgamated zinc rod which is immersed in dilute sulfuric acid; the two solutions are separated by a porous partition. The cell reaction is $Zn + Cu^{2+} — Zn^{2+} + Cu$ This yields a fairly constant electromotive force of about 1.08 volt; the internal resistance is a few ohm.

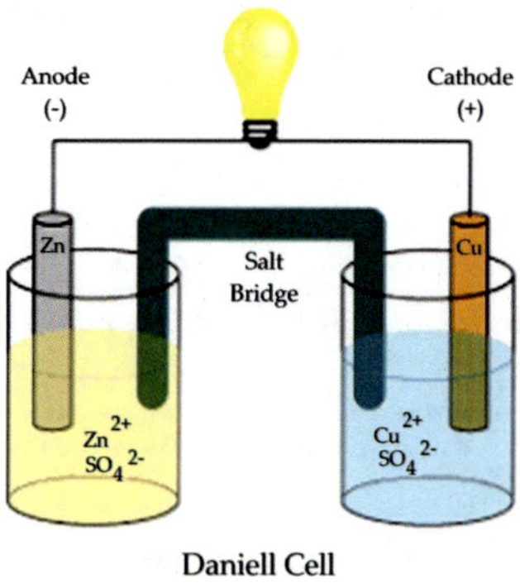

Figure 12. Daniell cell.

Source: Image by Wikimedia commons.

Dark Field Illumination

A type of illumination sometimes used on a microscope specimen to enhance visibility. The specimen is illuminated by light from the side rather than from underneath so that only light diffracted by the object reaches the eye. The technique is only applicable to the examination of small particles or fine lines, which appear as bright images against a dark background.

Dark Space

A comparatively nonluminous portion of an electrical discharge through a gas.

Dasymeter

An instrument for measuring gas density by finding the up thrust on a tube of known volume.

Davy Lamp

A lamp whose flame burns inside a wire mesh cage which cools the hot gases from the flame by conduction, thus preventing ignition by the flame of flammable gases outside the lamp. It was the first lamp which was safe for use in mines. It has now been replaced and is no longer used.

Day

The time taken by the Earth to rotate once around its axis.

Dead Room

A highly sound-absorbent room. It is made by covering all surfaces with sound-proofing material, structured so as to minimize sound reflection, and by placing absorbent screens at suitable locations so as to reduce the probability of setting up standing waves.

Dead Time

The post-stimulation time during which an electrical device is insensitive to another stimulus.

De Broglie Wave

A wave of wavelength hi (mv), associated with a particle of mass m moving with speed v; h is the plank constant. Since the square of the amplitude of the wave at a given point represents the probability of finding the particle at that point, the wave is sometimes regarded as a wave of probability.

Decay

The spontaneous disintegration of a radioactive nuclide into a daughter product, which may or may not be radioactive.

Decibel

A logarithmic unit used to compare two power levels, usually of sound or electricity. Power levels P and P0 are said to differ by n decibel where n = 10 log(P/Po). For sound,P0is usually chosen to be at the threshold of AUDIBILITY. One decibel is roughly the smallest detectable change, which explains why the decibel is preferred to the bel, which is ten times as large.

Declination

The angle between the horizontal component of the Earth's magnetic field at a point and the direction of true north at that point.

Defect

A discontinuity in the regular structure of a crystal. If only one single atom or molecule is involved, the defect is known as a point deject; three possibilities are illustrated in fig. D2. A Frenkel defect consists of a vacancy interstitial pair, formed when an atom moves to an interstitial position, leaving behind a vacancy.

A Schottky defect occurs either when an atom moves to the surface leaving a vacancy, or when an atom moves from the surface to an interstitial position; surface includes grain boundaries or dislocations. Point defect formation is encouraged by heat treatment, the number of defects rising exponentially with temperature. Strain and irradiation with ionizing radiation also result in point-defect production.

Degaussing

The process of neutralizing the magnetic field of an object, for example by applying an equal and opposite field produced by coils of wire carrying a current. Important applications are to ships in order to prevent them from detonating magnetic mines, and to television receivers in order to neutralize the Earth's magnetic field and so improve picture quality.

Degeneracy

The existence of an atom, electron or nucleus in more than one possible state corresponding to a given energy level. Degeneracy can be removed by applying a perturbation which results in a separation of the levels for the various states. The existence in classical dynamics of two or more independent modes of vibration having the same frequency. The state of a gas at temperatures below 20 K, for which the molecular heat is less than 3R/2 where R is the universal gas constant. The existence of atoms stripped of all their electrons. This occurs at very high densities, for example in neutron stars.

Degree

A unit of angular measure, it is also a unit of temperature difference. It simply means the power to which a variable is raised. If several variables are multiplied together, the overall degree is the sum of the powers; if the variables are added, the degree is that of the term of highest degree. Thus, y^4 is of the fourth degree while y^4z^3 is of the seventh degree, as is $y^4 + y^4z^3$.

Degrees of Freedom

The minimum number of co-ordinates required to specify the motion of each particle in a dynamical system. Any particle has three translational degrees of freedom, corresponding to motion in three directions. Molecules have, in addition, rotational and vibrational degrees of freedom.

A linear molecule has two rotational degrees of freedom, corresponding to rotation about two axes perpendicular to its axis; other molecules have three rotational degrees of freedom. The number of vibrational degrees of freedom of a molecule depends on the number of its atoms. By the principle of equipartition of energy, the energy for each degree of freedom is $kT/2$ where T is the absolute temperature and k the Boltzmann constant.

Delocalization

The existence in some molecules of electrons which are not confined to the region of one particular bond or atom but can move over the whole molecule. The molecule is then more stable than it would be if the electrons were localized.

Demagnetization

The removal of magnetism or magnetism properties. It is best achieved by taking the magnetized material through a series of diminishing hysteresis loops to zero. To do this, the material is placed in a coil of many turns which is connected to an alternating-current supply; the material is then slowly withdrawn.

Demand Pacemaker

A pacemaker which only stimulates a heart which fails to beat within a fixed period. The danger of ventricular fibrillation (inherent in ordinary pacemakers) should the heart return to normal rhythm is thus avoided.

Densitometer

An instrument for measuring the reflection density or the transmission density of a material. The most common form of the apparatus has a standard light source for illuminating the material and a photocell, connected to a microammeter, for measuring the radiation. Readings are taken with and without the specimen in place and the density calculated. An important use is the measurement of density of photographs of spectral lines, X ray diffraction patterns etc., and so deducing the composition of the radiation giving rise to the phenomena.

Figure 13. Densitometer.

Source: Image by Wikimedia commons.

Density

It is a measure of mass per volume. The formula for density is d = M/V, where d is density, M is mass, and V is volume. Density is commonly expressed in units of grams per cubic centimeter.

Dependent Variable

A variable whose value can be calculated when that of another variable, the independent variable, is known. Thus, in $y = 7x^6 + 2x$, x is the independent and y is the dependent variable.

Depletion Layer

The narrow region separating the n- and p-type components of semiconductor rectifier. It is denuded of charge carriers and a potential difference acts across it

Depolarization

The lessening of polarization effects in electrochemical cells. It may be achieved in a variety of ways: increasing the operating temperature, stirring the electrolyte, or chemically as in the Leclanche cell.

Depression of Freezing Point

The lowering of the freezing point of a solvent which occurs when a substance is dissolved in it. Relative molecular masses can be determined from the freezing point depressions produced.

Depth of Field

The range of axial distance on either side of an object whose image is in focus in an optical instrument, within which other objects also appear in acceptable focus.

Depth of Focus

The conjugate axial distance to the depth of field, i.e., the range of axial distance on either side of an image over which the image remains in acceptable focus.

Desaturated Color

A color whose chromaticity co-ordinates lie within the boundaries of the chromaticity chart; the closer the co-ordinates are to those of the white point, the more desaturated the color.

Detached Retina

A condition due to disease, injury or degenerative changes in the eye. If the retina is not promptly reattached to the choroid, blindness will result in the affected eye since alterations in the nutrition and metabolism of the retina when detached result in rapid and irreversible degeneration. A one millisecond flash of light from a laser focused on the retina will efficiently weld the retina back to the choroid; this procedure can be performed without anesthetic and without mechanical fixing of the eye and so is suitable for use in the outpatients' department.

Deviation

The difference between a particular measurement of a quantity and the mean value of all the measurements of the quantity. The mean deviation is the average of the magnitudes of all the deviations in a set of measurements. The standard deviation is the square root of the average of the squares of all the deviations in a set of measurements.

Dew Point

The temperature below which moisture from the atmosphere condenses on a surface in contact with it, i.e., it is the temperature at which the air is saturated and depends on the humidity.

Dextrorotatory

Denoting a substance which rotates the plane of polarization of polarized light in a clockwise direction, as viewed by an observer looking in the direction opposite to that of incidence of the light.

Diaphragm

An aperture, sometimes variable, whose center lies on the axis of an optical system and whose plane is perpendicular to this axis. The purpose of the diaphragm is to control the amount of light passing through the optical system.

Dichroism

The property of exhibiting two colors, usually one color by transmitted light and the other by reflected light. The property of some materials which exhibit double refraction whereby the ordinary and extraordinary rays are absorbed to different extents.

Dichromat

A color-deficient person all of whose sensations of color can be stimulated by a mixture, in suitable proportions, of only two primary colors.. There are three types of dichromat: one whose members require red and green primary lights; another whose members require red and blue primaries and the third whose members require green and blue primary lights; the last type is least common.

Dichromate Cell

A primary CELL for which poles of carbon and amalgamated zinc are immersed in a solution of potassium dichromate in dilute sulfuric acid. The electromotive force is 2.03 volts.

E

Earth

The third nearest planet to the Sun, situated a mean distance of 149,598,000 kilometers from it and moving round it in approximately one YEAR Earth's mass, diameter, relative density and period of revolution on its axis are about 6 x 10^{24} kilogram, 12,600 kilometers, 5.6 and 24 hours, respectively.

Figure 14. Earth.

Source: Image by Pixabay.

Echelon Grating

A type of *Diffraction Grating* made by stacking between 20 and 40 accurately parallel glass plates, of equal thickness to within a fraction of a wavelength, in optical contact to form a series of equal steps of about 1 millimeter width. The device may be used as either a reflection or transmission grating and will give high *Resolving Power* of between 10^5 and 10^6. The device is mainly used to study the fine structure of spectral lines.

Echo

A sound repeated as a result of the reflection of sound waves from a surface. It occurs for intervals between the sound and its reflection greater than 0.1 second. A high-pitched sound usually gives a better echo than one of low frequency.

Echo Sounding

A technique for measuring the depth of water by transmitting a pulse of sound into it and timing the arrival of the reflected pulse.

Eclipse

The temporary blocking of the light from the Sun or Moon, known respectively as a solar eclipse and a lunar eclipse, as viewed from Earth. An eclipse occurs when Sun, Moon and Earth lie in or nearly in a straight line.

Eddy Current

The current induced in a stationary conductor situated in a changing magnetic field, or in a moving conductor in a fixed field. Such currents occur in the cores of alternating current machinery, such as transformers, producing a loss of useful energy. It is usual to reduce these losses by lamination of the core or by using a ferrite core. Eddy currents produced in a moving coil instrument are often desirable because of the resulting damping, known as electromagnetic damping.

Efficiency

The ratio of the external work performed by a heat engine to the heat taken in, both quantities being measured in the same units. For a reversible heat engine, the efficiency is given by $(T_1 - T_2)/T_1$ where T_1 and T_2 are respectively the temperatures at which heat is taken in and given out. For real engines the efficiency is always less than this value.

Effusion

The leakage of gas through a fine orifice due to thermal motion of the gas molecules. When the mean free path of the molecules is small compared with the dimensions of the orifice, i.e., at ordinary pressures, Graham's Law applies. When the mean free path is greater than the dimensions of the orifice, i.e., at low pressures, the effusion rate is given by $s(kT/2nm)^{1/2}$.

Eightfold Way

A method of grouping the elementary particles which reveals regularities in their properties. It is the forerunner of Quark models.

Einstein

A unit of energy of electromagnetic radiation of a particular frequency, equal to the energy of one mole of photons of that frequency. Its value is N_A.h.v where N_A is the 'Avogadro constant,' h the 'Planck Constant' and v the frequency.

Einthoven Galvanometer

An instrument consisting of a single conducting filament stretched midway between the poles of a powerful electromagnet The deflection of the filament, which occurs when a current is passed through it, is measured by means of a high-power microscope let through one of the pole pieces. The galvanometer will detect currents as low as 10–11 ampere.

Elasticity

The property which enables a body, deformed by an applied load, to recover its original configuration when the load is removed.

Elastic Limit

The stress value above which a body no longer deforms elastically, i.e., the stress value at which plastic deformation starts.

Elastic Modulus

The ratio of stress to the resulting strain before the elastic limit is reached. There are different moduli corresponding to the various types of strain: examples are bulk modulus, rigidity modulus and young's modulus. For substances not obeying Hooke's Law, for example cast metal, concrete, marble and wood, the moduli are defined at a particular stress value as the ratio of a small increase in that stress to the resulting strain increase.

Electret

A dielectric substance which retains an electric moment after removal of the applied electric field. It is thus the electrical analog of a permanent magnet.

Electrical Calorimetry

A method of calorimetry, in which the rise of temperature of a known mass of substance, produced by supplying a measured amount of electric energy, is measured. The method is used for measuring specific heat capacity and specific latent heat.

Electric Displacement

A vector whose direction at a given point is that of the electric field at that point and whose magnitude is the produce of the magnitude of the electric field strength.

Electric Energy

Energy possessed by a charged body due to its position in an electric field. It equals QV, where Q is the body's charge and V the potential at the position of the body.

Electric Field

The space surrounding an electric charge and throughout which the charge will exert a detectable force on another electric charge.

Electric Field of Charged Conductor

Inside the conductor the field is zero. For a charged spherical conductor, the field outside the conductor is the same as if the whole charge of the sphere was concentrated at its center.

Electric Field Strength

The force experienced by a stationary unit positive charge placed in an electric field at a point where its strength is to be determined.

Electric Flux

The product of the area under consideration and the component, normal to this area, of the average electric flux density over it. It is therefore the surface integral of the electric displacement normal to the surface.

Electricity

The branch of physics concerned with the study of phenomena caused by electric charges. It is known as current when concerned with moving charges and as static electricity when concerned with stationary charges.

Electric Motor

A quiet clean device for converting electrical energy to mechanical energy (efficiency 75–95%), using electromagnetic induction. An electric motor may be considered as a generator which is supplied with current through its coil. Several types of design are available

Electrocardiography

The recording of potentials which have spread to the body surface from the heart. Since the potentials are small they are amplified prior to recording. To minimize the presence of nerve and muscle action potentials on the recording, it is made with the patient lying down and as relaxed as possible. The various direct-current potentials, which are of the order of 1 millivolt, are electronically eliminated from the recording. Studies of recordings enable heart disease to be diagnosed early and relatively simply.

Electrochemical Equivalent

The mass of a given substance which is liberated or deposited from a solution containing it at an electrode by the passage of 1 coulomb of electricity.

Electrochemistry

The study of electrolysis, electrolytic cells and the general behavior of ions in solutions.

Electrode

A conductor by means of which current is passed into or removed from an electrical system. The positive electrode is the anode and the negative is the cathode.

Electrodeposition

The coating of one metal on the surface of another metal: the metals are made the electrodes in an electrolytic cell which contains a solution of a salt of the metal to be deposited. The thickness of the deposited layer is governed by the quantity of electricity passed.

Electrode Potential

The potential difference between an electrode and the electrolyte with which it is in contact.

Electrodynamics

The study of the mechanical forces existing between neighboring current-carrying conductors.

Electrodynamometer

An electric measuring instrument using the force between two or more current-carrying coils. With appropriate connections it may be used as a voltmeter, ammeter or wattmeter for either alternating or direct current.

Electroencephalography

The study of electric potentials on the surface of the head. The potentials are of the order of 100 microvolt. Electrodes are hooked to the scalp at up to 24 locations and the pattern of potential difference between any pair of electrodes can be studied. Prior to recording, background noise and other electrical disturbances in the body are electronically filtered out of the signals; the signals are then amplified and recorded.

Electrolysis

The passage of an electric current through a liquid containing ions. The current flows in both directions and so differs fundamentally from the electron flow in a metal conductor. An important aspect of the subject is chemical change occurring at the electrodes.

Electrolyte

A solution of a substance, or the substance itself in a fused state, which conducts electricity by the movement of positive and negative ions.

Electrolytic Gas

The gas mixture produced by the electrolysis of water, i.e., an explosive mixture of hydrogen and oxygen in the proportions 2:1 by volume.

Electrolytic Separation

The separation of isotopes by making use of the different rates at which they are released in electrolysis. For example, in the electrolysis of water less deuterium than hydrogen prorate is formed at the cathode since deuterium is heavier than hydrogen; in time therefore, the water becomes enriched with heavy water.

Electromagnet

A coil of insulated wire wound round a soft ferromagnetic core: when current passes through the coil the core is strongly magnetized but it loses its magnetism when the current is switched off. Electromagnets are used for example in electric bells, switches, solenoids and lifting cranes for some metals.

Electromagnetic Field

A field representing the interaction of electric and magnetic forces. For free space, Maxwell's equations, express the relationships between the variables; the pointing vector gives the direction and magnitude of the energy flow. According to electromagnetic field theory, an electric field is set up by stationary electric charges and a magnetic field by moving electric charges. The nature of the observed field will therefore depend on the observer's relative motion, so that a field appearing magnetic to one observer may appear electric to another and vice versa.

Electromagnetic Focusing

The use of a magnetic field to produce convergence of an electron beam. The field is formed by passing direct current through a coil which is usually short and coaxial with the beam.

Electromagnetic Induction

The production of an electromotive force in a conductor by changing the magnetic flux linked with it. The flux can be varied either by varying the magnetic field or by altering the configuration of a conductor relative to a constant magnetic field: the first method is used to produce the current in the secondary coil of a transformer; the second method is employed in producing current in a generator. According to the Faraday-Neumann law, the magnitude of the induced electromotive force is proportional to the rate of cutting of the magnetic flux. According to Lenz's law, the force is in such a direction as to oppose the change. The combined statement of the two laws is V = –L dl/dt.

Electromagnetic Interaction

A type of interaction between elementary particles arising from any electric and magnetic fields associated with the particles. Electromagnetic interaction has about l/200th of the strength of strong interaction. In an electromagnetic interaction there is conservation of angular momentum, baryon momentum, charge, charge conjugation parity, isospin parity and strangeness.

Electromagnetic Pump

A simple pump without moving parts and suitable for use with conducting liquids. A strong magnetic field is applied across a diameter of a pipe holding the liquid, through which a current is passed. The liquid therefore experiences an induced force acting along the axis of the pipe and is thus propelled along the pipe. An important application of the pump is the removal of the liquid-sodium coolant used in some nuclear reactors.

Electromagnetic Radiation

Waves of energy caused by the acceleration of charged particles. The waves involve transversely oscillating electric and magnetic fields at right angles to each other. The waves travel through a vacuum at the speed of light and are described by Maxwell's equations.

Electrometer

An instrument for detecting or measuring potential difference. Originally electrostatic instruments based on the electroscope, such as the electrostatic voltmeter and quadrant electrometer, were used. Apart from special applications these have now been superseded by electronic devices such as an amplifier with very high input impedance drawing negligible current and employing field effect transistors. Electrometers can be used to measure currents as low as 10^{-9} ampere by passing the current through a known high resistance and measuring the resulting voltage drop with the electrometer.

Figure 15. Electrometer.

Source: Image by Wikimedia commons.

Electromotive Series

A series of the metals arranged according to their readiness to form positive ions. The more electropositive an element, the greater its ability to replace less electropositive elements from their salts.

Electromyography

The process of registering the electrical activity of muscle. Needle electrodes are inserted into the muscle under investigation and their output is either displayed on a cathode ray tube, or fed to a loudspeaker. The technique is considered to be an efficient means of diagnosing disease of the lower motor neurons, myoneural junctions and skeletal muscle fibers.

Electron

An elementary particle of negative charge, the magnitude of which is 1.602192×10^{-19} Coulomb. It has a mass equal to 9.10956×10^{-31} kilogram. It has a SPIN of it and is classified as a lepton. Electrons are constituents of all atoms and, when free, are primarily responsible for electric conduction in most substances. Electrons are produced in beta decay, in electric discharges and in thermionic and photoelectric emission. The antiparticle of an electron is the positron.

Electron Affinity

The energy released when an electron is attached to an atom or molecule. It is frequently positive, i.e., the negative ion is often more stable than the neutral species.

Electron Beam Technology

The use of electron beams in a variety of applications. These include the refining, casting and welding of refractory metals, for example tungsten; hole drilling and milling of both metals and ceramics; treatment of deep-seated malignant tumors.

Electron Bombardment

In interaction with matter, an electron may undergo either elastic or inelastic collisions. The former results in deflection of the electron and dissipation of some of its energy in the form of heat: this is the basis of the localized heating of substances in a vacuum. In the case of inelastic collision, the electron loses discrete amounts of energy, accompanied by various phenomena according to the state of the matter: for gas atoms, ionization or excitation accompanied by light emission occurs; for solids and liquids, emission of X-rays and fluorescence and secondary electron emission are among the phenomena which may result; for crystalline matter, electron diffraction might occur.

Electron Density

In general, it refers to the number of electrons per unit mass whereas in astrophysics, it refers to the number of electrons per unit volume.

Electron Density Map

A representation of the location of atoms in a crystal lattice as deduced from X Ray Diffraction studies. The atoms do not appear as small point structures since the electrons are not localized; thermal vibrations of the atoms also have an effect. With the help of such representations, the structure of many compounds which defied chemical analysis has been determined.

Electron Diffraction

The diffractions of electrons caused by atoms or molecules. The fact that it occurs confirms the wave nature of the electron, the wavelength being h/

(mv) where h is the Planck constant and m and v the mass and speed of the electron. The penetrating power of an electron beam is low compared to that of an X ray beam and so electron diffraction is not much used to investigate crystal structure. It is widely used, however, for solid surface studies and for the measurement of bond lengths and angles of molecules in gases. The main techniques are low-energy electron diffraction, known as 'leed,' in which the diffracted electrons are reflected on to a fluorescent screen, and high-energy electron diffraction, known as 'heed,' in which electrons reflected from or transmitted through thin films are studied.

Electron Gas

A population of free electrons within a vacuum or a metallic conductor. It is considered similar to a real gas dissolved within a substance. The model is applicable and beneficial in the theoretical treatment of for example conductivity and thermionic emission.

Electron Gun

An electron gun, also called an electron emitter, is a device, generally used to deflect a beam of high velocity electrons, for example, in a television picture tube. The electron beam passing through the control grid is managed by adjusting the grid potential as per the intensity of the beam.

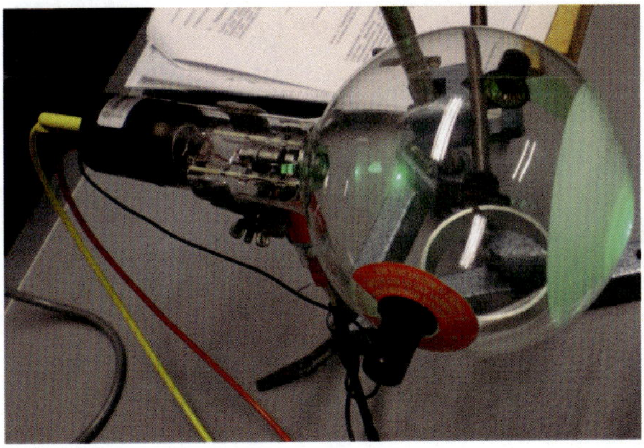

Figure 16. Electron gun.

Source: Image by Wikimedia commons.

Electronics

The branch of physics, that caters to the analysis, design and usage of devices depending upon electron movement within a gas, vacuum or semiconductor. Gas and vacuum appliances have very limited specific uses and are becoming obsolete.

Electron Impact Spectroscopy

An inclusive term for forms of electron spectroscopy in which excited molecular states are studied by using electron beams to induce transitions between electronic energy levels.

Electron Lens

A device or lens that either converges or diverges a beam of electrons by employing either an electric or magnetic field, in an analogous manner to the light beam in an optical microscope. Electron lenses are used in objects with sharply focused electron beams, for example, cathode-ray tubes, television camera tubes and electron microscopes.

Electron Microscope

An instrument that obtains images with higher resolution than those in an optical microscope. It is used to produce magnified images of a specimen by using a beam of energy electrons instead of a light. The electron wavelength is much shorter than that of light and so the resolution of the electron microscope is much better than that of its optical counterpart. Objects as small as 0.2 nanometer can be resolved and a magnification of 200,000 times obtained.

F

Faced-Centered

Containing or incorporating a crystal structure, where the UNIT CELL possesses eight atoms at the edges and an additional six atoms located one at the middle of each face. The lattice is explained in the form of a face-centered cubic, if the unit cell is a cube.

Fahrenheit Scale.

Refers to a temperature scale where the ice point is explained as 32°F as well as the steam point as 212° F. The scale is no more implemented in scientific purposes.

Fallout

Radioactive substance settling up on the surface of the earth as an outcome of a nuclear explosion. Local fallout takes place enclosed in 250 kilometer of an explosion and occurs within a couple of hours of it. Tropospheric fallout takes place in every part of the Earth in the estimated latitude of the explosion and under a week of it. Stratospheric fallout might take place anywhere on Earth over a duration of time in wake of the explosion. Iodine–131, which assembles in the thyroid gland, as well as strontium–90, which gathers in bones, are the highly detrimental FISSION fractions since they might be passed to humans as they are taken up by grazing animals and through this way to living beings.

Farad

Symbol F. The SI unit of CAPACITANCE, equivalent to the capacitance of a parallel plate capacitor which holds up a charge of one coulomb when there exists a possible variation of one volt throughout its plates. The microfarad proves to be more beneficial practical unit.

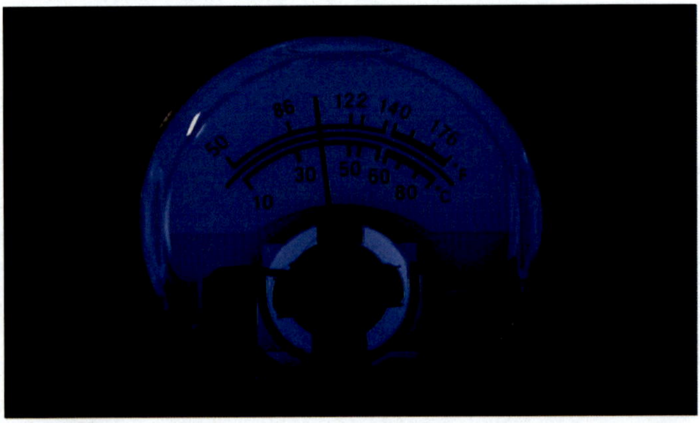

Figure 17. Farad.

Source: Image by Pxfuel.

Faraday Effect

The spinning of the plane of polarization of the radiation of plane-polarized electromagnetic on moving through specific isotropic material subject attached to a magnetic field, strength H, collateral to the regulation of progression of the radiation. The angle of rotation is in proportion to Hl, where l stand for the length of the path of the radiation that is present in the medium. If the radiation again passes over its path in the reversed direction the result of rotation is intensifies, that is, the direction of rotation is individualistic of the manner in which the radiation travels throughout the magnetic field.

Fatigue

The continuous reduction of the *breaking stress* of a substance subjected to reiterated applications of a stress which, if it had been sustained at the similar steady value, would not have given rise to unsuccessful outcomes.

Feedback

A procedure in which the product regulates the input in some manner, for instance the integrating of a fraction of the product of an electronic detector to the input. If in case, the input is magnified through the process it is called as positive feedback, on the other hand, it refers to as the negative response if the input is lowered.

Fermi

A unit of length which is equivalent to 10^{-15} meter, largely implemented in nuclear as well as atomic physics.

Ferrimagnetism

A phenomenon representing *antiferromagnetism* but is linked with magnetic moments that are disparate adjacent and antiparallel.

Ferrite

An enveloping low-density ceramic oxide of iron along with another additional oxide, which demonstrates whether or not the ferrite displays *ferrimagnetism* or *ferromagnetism*.

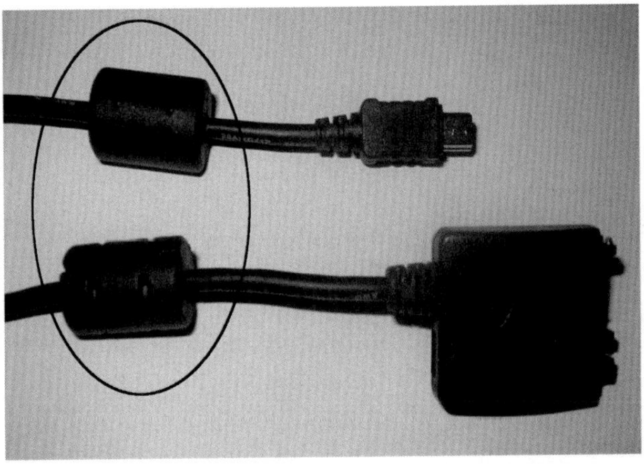

Figure 18. Ferrite.

Source: Image by Wikimedia commons.

Ferroelectricity

The characteristic, displayed by a few dielectrics like barium titanate, of holding onto the electric *polarization* without the polarizing field. The result is because of the arranging done in dimensions of electric dipoles located in the dielectric in a symmetrical manner to magnetic dipoles in ferromagnetic substance (refer *ferromagnetism*). The electric displacement graph in opposition to the implemented electric field displays a *hysteresis* loop which is similar to that of ferromagnetic hysteresis.

Ferromagnetism

A kind of magnetism demonstrated by a few solids like iron, cobalt together with nickel that become magnetized in feeble magnetic fields and which possess a higher positive *magnetic susceptibility*. These types of substances show *hysteresis* and have extreme relative *permeability*. The magnetic conduct of ferromagnetic material is based on temperature.

A few of the ferromagnetic substances for example, steel holds back ample of their magnetism at the time of the withdrawal of magnetizing field, and are considered to be hard; while the rest others, for instance, soft iron, lose track of the nearly all of their magnetism on elimination of the field and are considered to be soft.

Fibonacci Numbers

A pattern of numbers every one of which is the aggregate of the two previous numbers: 1, 1, 2, 3, 5, 8, 13.

Fiber Optics

An optical method used for passing on images together with transparent fibers that are mobile. On the end of a glass fiber having high refractive index, the light which falls is less than the diameter of 1 millimeter, moves through by reiterating *total internal reflection* and thus goes through a very less amount of absorption or completely no absorption at all. By making the use of a bundle of these fibers (each having the diameter of 0.01 millimeter to 0.5 millimeter) in a riveted array, absolute images can be transferred, each fiber being accountable for a little domain of the whole. The fact that the total reflection on the inside can take place even if the fibers are considerably curved, the framework is of immense practical importance for looking over or photographing objects that are not easily accessible, for example, the parts of the machine together with internal organs of the body.

Field

An area under the impact of a few of the physical agency, for instance electrical, magnetic, gravitational, and thermal; where each of the first three is referred to as *vector* whereas the fourth is said to be *scalar*. A vector domain is often identified through a set of curves, each is regarded as a line of force, which possess density at any given point provides the energy of the region at that extremity and its direction is the direction of that area at the point.

Field Effect Transistor

A *transistor* inside of which current flow relies on the tendency of most of the carriers only. There are two predominant types: junction region impact transistors, written JUGFET or JFET, secondly, the insulated gate region impact transistors, written IGFET; the second one is also stated by the abbreviations like MISFET or MIST or MOSFET or MOST. For both the kinds the electrodes become the SOURCE, GATE as well as DRAIN. The fundamental structures are demonstrated in fig. F2. For the n type medium JFET, responsible for making the drain positive comparative to the origin causes electrons in order to flow from origins to drain.

Field Emission

The release of electrons which is made from the exterior of a solid because of the high electric field which is present at the surface. The use of the phenomenon is applied in the field emission microscope in order to analyze the surface framework of a metal point which is located in the middle of a screen which is said to be hemispherical fluorescent screen; the screen is set up at an extreme positive prospective in context to the point. The whole is circumscribed in gas at decreased pressure. A visual of the point is constructed on the fluorescent screen through the discharged electrons.

Field Ionization

The ionization of gaseous atoms together with the molecules on the solid's surface under the impact of an excessive electric field: the movement of electron takes place from either a gaseous atom or molecule to the solid giving rise to a positive gaseous ion, which is quickened away through the field. The eventual outcome is applied in the field ion microscope, which is identical to the FIELD EMISSION microscope excluding that the fluorescent screen is negative in context of the metal point and the visual is because of the effect of gas ions present on the screen. The device contains high resolution.

Film Badge

In a plastic holder, a fraction of photographic film, exhausted by those revealed to radiation. The darkening of the film signifies the amount of exposure. If the obtained amount of a specific radiation is of benefit, filters to eliminate other kinds of radiation are positioned ahead of the film.

Filter

An instrument for capturing radiation of specific energies in the course of transferring radiation of various other energies. For instance, light may be filtrated by fractions of glass or film that are colored. An electric circuit which enables channels of alternating currents enclosed in a certain domain or ranges of frequency, at the time of stopping currents which possess frequencies exterior to the allowed range(s).

Filter Pump

Refers to the type of vacuum pump, in which a jet of water is enforced from starting to end, in a tapered nozzle which captures air molecules and eliminates them from the system. It does not have the ability to lessen the pressure below the water's vapor pressure.

Finder

A small telescope associated to the margins of an immense astronomical telescope in order to make possible the regulation of the big telescope with regard to the object so as to be discovered.

Fine Beam Tube

A gas discharge tube having low-pressure in which the channel of a magnificent beam of electrons supplying by the medium of a hole in the anode is exhibited evident because of the ionization of the hydrogen molecules. By implementing a pair of *Helmholtz coils* to use in the form of a uniform magnetic field to the tube, and estimating the end result of the electron diversion, the tube might also be applied when it comes to the evaluation of elm for the electron.

Fine Structure

The structure which is discovered in either a spectral line or band at the time when it is perceived at high resolution; thus, splitting of line due to the spinning of electron can be discovered. In order to determine *hyperfine structure* even inflated resolution is needed. Hyperfine structure is the consequence of energy-level transformations in atoms of various isotopes of a material and also from associations of nuclear spin together with electron spin.

Fission

The procedure in which an atomic nucleus breaks down into a fraction nucleus of relative size. It most commonly goes along with emission of neutrons together with gamma rays and the liberation of abundance of energy. Fission might be either rapid or instigated by shelling. Compare *nuclear reactor* and *nuclear weapon*.

Faxed Star

A star which emerges not to change its location on the *celestial sphere*. A faxed star was given such a name in order to differentiate it from a planet, which was referred to as a wandering star.

Flashover

An unnecessary arc release that takes place between two electrical conductors. Destructions can be reduced by making the use of a screen of fireproof substance, that is a flash barrier.

Flash Point

The minimum temperature at which a material must be given heat prior to when it can be ignited.

Flavor

The general name given to the qualities which differentiate the different types of QUARKS, and along with this the different types of LEPTONS as well.

Flip-Flop

An electronic circuit which contains two stable phases and is transformed from one phase to the other through an activating pulse. Such type of a circuit is extensively applied as a LOGIC CIRCUIT.

Floppy Disc

A mobile plastic DISC which comes in addition to a magnetic coating, held in an inelastic covering. It is applied for reserving and keeping the information in a small COMPUTER system.

Flow

The flow of an ideal liquid is the stable streamline flow throughout a tube of an ideal liquid is controlled by the BERNOULLI EQUATION. In addition to this, since the liquid is free of friction, its speed remains similar at every point on a provided cross section and, as a matter of fact, the liquid is dense, the outcome of speed of flow as well as cross-sectional region at any given point is uniform, that is, the slender the tube the more fast the liquid moves through it. The fact that a real liquid possesses some internal friction due to which there is a decline in pressure throughout a tube in which it moves through, denoting that energy is being dissolute. Moreover, the speed of a substance of liquid present in a tube relies upon the length of the molecule from the tube axis.

Fluidics

The flow of fluid applied through slender pipes so as to carry out tasks generally or prior to this performed by electronic equipment. The circuits of fluid are way too slower as compared to the electronic ones but remains unaltered by magnetic fields or radiation and are less responsive to temperature. Thus, they have growing significance in NUCLEAR REACTORS as well as SPACECRAFT.

Fluorescence

Luminescence which terminates during the time when the exciting source is disconnected. Fluorescence is a term which is applied scientifically in order to elucidate luminescence remaining less than 10^{-8} second after elimination of excitation, or luminescence which emerges to the eye to terminate with excitation. In common usage the term is often identical with luminescence.

Fluorescence Assay

A technique applied on wider scale for the recognition and location analysis. It is sometimes act as a substitute to radioactive methods since it is free of hazard and has the ability of identifying contamination at a concentration of around 1 in 1010• Monochromatic radiation is applied in order to excite the FLUORESCENCE and allocation is made for any derived fluorescence of vessels, solvent etc. For a few works it is important to initiate the availability of fluorescence; in more defined investigations the fluorescent light is scrutinized and explored by SPECTROMETER.

Fluorescent Lamp

A lamp in which the light is predominantly generated by LUMINESCENCE. Most commonly, the lamp comprises of a long glass GAS DISCHARGE TUBE in which the inner surface is laminated with a PHOSPHOR; the gas is generally mercury vapor in which, when current moves along, it releases ultraviolet radiation which sequentially stimulates evident radiation from the phosphor covering. These types of lamps transform excessive of the energy dispensed into light as compared to the tungsten filament lamps.

Fluxmeter

An instrument used for estimating MAGNETIC FLUX. In one kind an EXPLORING COIL associated to a BALLISTIC GALVANOMETER is randomly initiated into the variability with its plane perpendicular added to it. The galvanometer reading is in proportion to the MAGNETIC INDUCTION.

Flying Spot Microscope

A microscope in which a little spot of light examines the object. The visual is obtained on a photocell, and the emerging output is intensified and sustained to a cathode ray oscillograph in which the time base is coincided with the spot of light. The comparison of the display might differ and the resolution is superior as compared to the one that is received by PHOTOMICROGRAPHY.

F-Number

In a lens, the ratio of the focal length to its diameter, is the F-number. In a photographic camera, various f numbers for the identical lens are acquired by applying a variable diaphragm to alter the constructive lens diameter. The brightness of the image diminishes with expanding f number and so extended exposure times are needed at extreme f numbers; the f number of a lens is most of the times known as the speed of the lens.

Focal Length

The length between a crucial point in an optical system and the correlating focal point. Normally, a system contains two focal lengths, that is f_1 and f_2, associated by the equation $n_1/f_1 = -n_2/f_2$ in which n_1 and n_2 are known as the REFRACTIVE INDICES of the media on the two borders of the system. For a narrow lens and paraxial rays, the chief points coexist at the middle of the lens; when it comes to a spherical mirror and paraxial rays, they clash with each other at the center point of the mirror.

Forbidden Transition

A transformation, either by an atom or molecule between two levels of energy which would yield in a QUANTUM NUMBER, where transition is not enabled through the SELECTION RULES.

Force

The agency which is accountable for changing the state of rest or of constant motion of a body. Some common instances of forces are Buoyancy, Centrifugal Force, Centripetal Force, Electromotive Force, Intermolecular Force, Saturation Force, as well as Viscous Force. The SI unit of force is called the NEWTON.

Forced Oscillation

The motion which generates when a vibrating system is performed upon by a vibrating force which is external. The system oscillates containing a frequency which is equivalent to that of the force which is external. The magnitude is at extreme when this frequency is identical as the natural frequency available in the system; RESONANCE is then said to take place.

Force Pump

A pump which is nearly identical to the LIFT PUMP but with a valve in the cylinder side instead of being in the piston. The force pump has the potential of elevating water from a depth which is not more than 100 meters.

Fractals

Geometric shapes which copy objects which emerges to be self-identical under various magnifications. For instance, the figures of coast lines, of polymers and mass dispensation in indiscriminate alloys. The notion is being progressively implemented in statistical physics.

Fraction

An expression in the shape of one quantity, the numerator, dissected by another, the denominator. If in case, both numerator and denominator are integers, the fraction is supposed to be a common fraction; if both numerator and denominator are themselves fractions, the fraction is supposed to be a multiplex fraction. A fraction if called as a proper fraction when the numerator is smaller as compared to the denominator; if the numerator is bigger in comparison to the denominator, the fraction is supposed to be an improper fraction.

Frame of Reference

A specific set of coordinate axes applied as a reference system in order to carry out physical estimations.

Fraunhofer Lines

Fine dark lines present in the solar spectrum, mostly caused by the integration of definite wavelengths in the Sun's PHOTOSPHERE. Some of the lines are because of the absorption in the atmosphere of the earth.

Free Electron

An electron that is temporarily connected to a particular atom or molecule and so is on the lose as it passes and shifts under the impact of an electric field which is implemented.

Free Oscillations

Oscillations that take place when a system is dislocated from its balanced equilibrium position and then left to be itself. The system carries out vibrations around its position at equilibrium together with its natural frequency, that is the attribute frequency of the system. The magnitude of oscillation decreases eventually until all the energy contribute in the displacement which occurs in the beginning and has been dissipated.

Frequency Distribution

A table, graph or either equation delineates how a specific attribute is dispensed amidst the members of a group. An instance of this is a plot of the number of people enclosed in a group which have heights lying in different specified durations into which the height span is split up. The number in every duration is known as the suitable frequency to that duration. Compare HISTOGRAM.

Frequency Modulation

MODULATION in which the transporter of frequency is expanded or decreased as the signal magnitude expands or diminishes, and at the available frequency of the signal, the carrier magnitude persists constant.

Friction

In the dynamic sliding friction. The force is required to hold one solid surface just coasting over another surface which is horizontal. It is equivalent to $\mu_d R$ where R refers to the normal reaction between the exteriors and μ_d is a fixed which is known as the coefficient of dynamic friction. Second type of fiction is the static sliding friction. The force mandatory to begin one solid surface just coasting over another horizontal surface.

It is equivalent to $\mu_s R$ where R stands for the normal reaction between the surfaces and μ_s refers to a constant own as the coefficient of static friction; μ_σ is to some extent greater as compared to μ_d, the coefficient of dynamic friction. Third one is the rolling friction. The force holding out against the rolling of a body on a horizontal plane. It is way too smaller as compared to the dynamic sliding friction, mainly for hard surfaces, as a matter of fact which elucidates the use of wheels along with the ball bearings.

Fringes

Circular or rectilinear or different other shaped specimen of color, or of another monochromatic brightness as well as darkness, which is generated by INTERFERENCE or DIFFRACTION of light.

Fuel Cell

Refers to an electrochemical cell inside of which the energy emerging from the reaction between a traditional fuel and oxygen, most commonly from the air, is directly and constantly transformed into energy which is of low-voltage and direct-current electric energy. Charging is not needed in this.

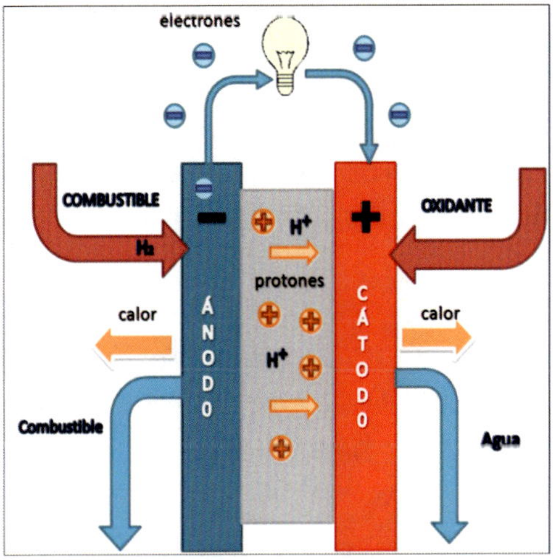

Figure 19. Fuel cell.

Source: Image by Wikimedia commons.

Fusion

A nuclear reaction whereby the two light nuclei associate so as to form a heavier nucleus, for instance, the formulation of helium out of hydrogen: ${}^{2}{}_{1}H + {}^{3}{}_{1}H\text{-+}\ {}^{4}{}_{2}He + {}^{1}{}_{0}n$. Such type of reactions goes along together with the evolution of immense amounts of energy. High temperature is required for the reaction fusion reactor 115 as it is only because of it that the kinetic energy of the nuclei is substantially great to control their repulsion. The procedure is the foundation of the formation of energy in stars and was first determined on Earth through the explosion of the hydrogen bomb. Attempts are being carried out to harness the method as a regulated energy source.

Fusion Reactor

A device used for acquiring energy through the process of nuclear FUSION. Generally, gas is at first heated through passing over a current through it, which gives rise to the construction of a PLASMA as an outcome. The plasma is restrained by appropriate magnetic fields. The difficulties of the firmness of plasma and the removal of its energy are still under examination.

G

Galaxy

The wheel shaped spiral Catherine galaxy having the Sun. The diameter of the Galaxy is approximately 105 light years and its width is approximately 20,000 light years. It comes under a local group of approximately 20 galaxies.

Galilean Telescope

A simple terrestrial TELESCOPE having a convex objective lens as well as a concave eyepiece distinguished by the variation between the focal lengths of the two lenses. Though the final image is erect, the system is applied in low-power opera glasses along with simple binoculars.

Figure 20. Galilean Telescope

Source: Image by Wikimedia commons

Galilean Transformation

A pattern of equations linking to space and time co-ordinates, i.e., x,' y,' z' and t,' in a structure of reference going with constant speed v through the x axis of different frame, to the co-ordinates x, y, z and t in this other structure. Every observer is anticipated to be at the beginning of co-ordinates of the accurate frame and to correspond at $t' = 0 = t$. The given equations· are

$x' = x - vt$, $y' = y$, $z' = z$, $t' = t$. They obey to Newtonian Mechanics.

Galton Whistle

Refers to a short cylindrical pipe blustered from an annular adaptable nozzle. For an appropriate air-blast pressure together with nozzle setting, resonant vibrations of frequency over the audible limit are acquired.

Galvanometer

Any instrument used for examining or estimating small electric currents. The most usual type is a moving coil instrument, which is, although, not so appropriate when it comes to alternating current.

Games Theory

A mathematical theory implied and used in order to anticipate the optimum strategy in circumstances of conflict. 'γ. The symbol implemented to signify the ratio of the heat of a gas at uniform volume to that at fixed pressure. Secondly, the descent of the linear fraction of the graph of density of prepared photographic material in opposition to logarithm of exposure.

Gamma Camera

A device for determining the dispensation of radioactive isotope in a section of the human body. A huge thin Scintillation crystal is detected above the section under examination and below an arrangement of Photomultiplier tubes. When gamma rays from the isotope attack the crystal at segment A say, the size of the eventual pulse from each photomultiplier relies on its position comparative to A By catering to the outputs to a cathode ray Oscillograph through an amplifier, an image of the isotope dispensation is acquired and can be captured in a camera.

Gamma Ray

Highly penetrating electromagnetic radiation of wavelength in the span w-Jo meter to 10–13 meters, that is containing the smallest wavelengths in the familiar electromagnetic spectrum and therefore the biggest photon energies. Gamma rays are radiated voluntarily through some radioactive substances in the course of nuclear decay and are also constructed in particle Annihilation. Dissimilar alpha and beta rays, gamma rays are uncharged and are thus not deflected through electric as well as magnetic fields.

Gamma Ray Astronomy

The analysis of gamma rays from space. Examinations have been made from numerous satellites that was launched from Earth and the initial sources of several gamma ray origin located.

Gamma Ray Heating

A technique of supplying a regulated quantity of heat to a thermally abandoned sample at a temperature which is below 1 K by the irradiation of gamma ray, commonly from Cobalt–60. The method is applied in decrease temperature calorimetry.

Gas

A state of matter distinguished by smooth flow, compressibility and rapid growth in order to fill any container in which it is kept.

Gas Breakdown

A phenomenon which takes place when the voltage throughout a gas-filled tube goes beyond a certain value. Then the ions in the gas become adequately energetic to generate more ions through collision along with neutral atoms and to dodge recombination. Therefore, multiplication takes place, causing breakdown of the gas insulation attribute. Gas breakdown is to some extent analogous to Avalanche Breakdown resulting in a semiconductor.

Gas Cooled Reactor

A thermal nuclear reactor which applies a gaseous coolant. Carbon dioxide emerging at the temperature of 350°C is implemented in the mark I MAGNOX reactor; in the progressed gas cooled reactor it appears at 600°C temperature.

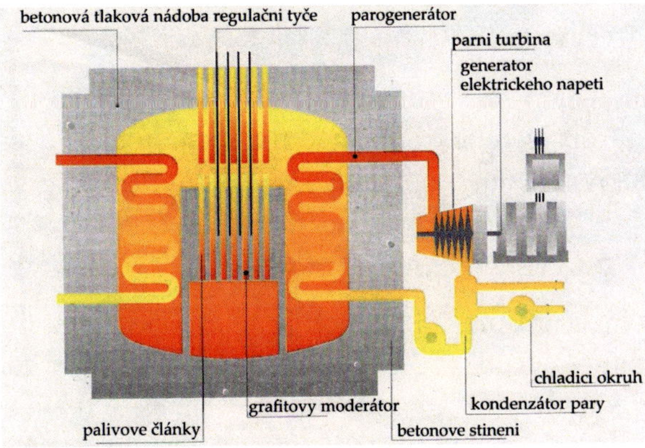

Figure 21. Gas cooled reactor.

Source: Image by Wikimedia commons.

Gas Discharge Tube

An ELECTRON TUBE whose property is remarkably impacted by the availability of a gas. For an adequately large electric field between the electrodes, ionization of gas atoms as well as molecules close to the electrodes takes place. As an ion shifts towards the electrode of opposite side to itself, clashing with other atoms and molecules might take place giving rise to their excitation or ionization and therefore the process extends. Radiation takes together both return to the unexcited phase as well as the recombination of ion.

Gaseous Ions

Ions produced in gases through the process of ionizing radiation. They spontaneously reunite in order to construct neutral atoms in addition to the molecules and therefore some of them remain shortly after dividing the ionizing radiation.

Gate

The electrode(s) present in a FIELD EFFECT TRANSISTOR which has a voltage that regulates the conductivity of the channel. Secondly, an electronic circuit which provides an output signal for only specific mingling of two or more input signals. Thirdly, an electronic device that travels through signals for only a particular fraction of the input signal.

Gauge Symmetry

A symmetry in which no estimated property of the world transforms if protons and neutrons can be exchanged for each other at every given point in space separately and unaccompanied.

Gaussian Eyepiece

An eyepiece which at times is applied in telescopes. The hole in the side of the eyepiece tube enables radiance of the crosswires; when they are the center of focus and seen juxtaposed with the visuals of them constructed by reflection in the plane mirror M, the telescope is concentrated for parallel light and have a perpendicular axis to M. Such an arrangement is significant for the accurate implementation of a spectrometer.

Gegenschein

A patch of light scarcely observed in the night sky and is placed opposite to that of the Sun. It is the outcome of the scattering of sunlight through meteor particles.

Geiger Counter

A device used for detecting ionizing radiation along with counting particles. It comprises of a cylindrical metal cathode having either halogen or rare gas at the condition of a low pressure, and at one end, it has a thin mica, glass or metal window; a thin wire anode is ascended across the axis of cylinder. The prospective variation between cathode and anode is a bit lower than that which is needed to generate a discharge in the gas, for halogen filled tubes that is between 300 and 400 volt and approximately 1000 volt for ones that are filled with rare gas.

Generation

A collection of two LEPTONS and two QUARKS. The first generation consists of electron, neutrino, up quark, down quark; whereas the second generation consists of muon, neutrino, charmed quark and strange quark; and the third generation consists of tau lepton, neutrino, hypothetical top quark as well as the bottom quark.

Geophysics

The research of the behavior as well as the physical properties of Earth. It incorporates geophysical inspection, seismology, oceanography, together with the study of atmospheric electricity, terrestrial magnetism and gravitational properties

Getter

A substance, generally a pure metal, implemented in the use for separating unwanted atoms or molecules from an environment. For instance, heated magnesium will amalgamate with unnecessary excess of oxygen and nitrogen inside a sealed vacuum system, and the use of phosphorous is done in the production of metal oxide semiconductors in order to eliminate unwanted sodium.

Glow Discharge

The end product of radiation across the length of a GAS DISCHARGE TUBE at an appropriate gas pressure. Characteristic traits of the dispensed glow are not stable difference of voltage with length across the tube, comparatively high current of a few milliampere, on current, the independence of voltage drop. Due to the latter characteristic the glow release proves to be a useful voltage stabilizer. Other applications are present in luminous signs, for lighting purposes in addition to this it is also a source for spectral scrutiny of either the gas or vapor in which the dispensation takes place.

Gluon

A hypothetical particle which is swapped between a pair of quarks and combining the pair jointly, is a gluon. The gluon plays a vital role in Quantum Chromodynamics parallel to the task performed by the photon in Quantum Electrodynamics.

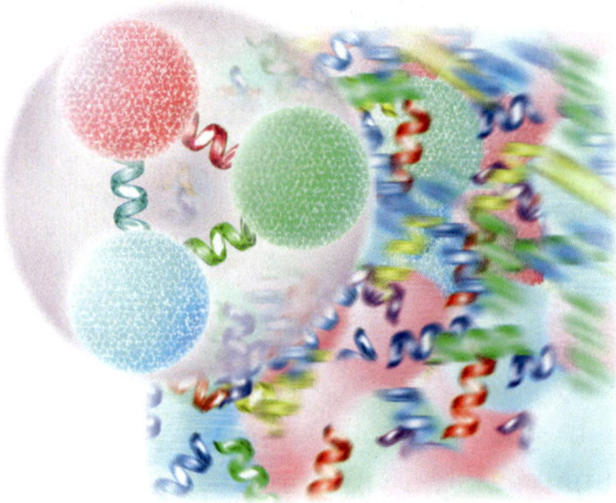

Figure 22. Gluon.

Source: Image by Wikimedia commons.

Golden Section

The splitting of a line or region so that the ratio of the whole to the larger fraction equals the ratio of the larger fraction to the smaller one. Proportions that are dependent upon golden fractions are regarded aesthetically agreeable and so emerge in several paintings.

Graticule

A chain of fine lines instigated into an optical device so that it may catch the sight simultaneously with the perceived image. It is applied as a reference system and also for carrying out estimation. The lines may be regarded as either fine wires or threads, or are imprinted on transparent material.

Gravimeter

A device used for the determination of fluid density by making the use of a total submersion float associated by a spiral spring to the bottom of the accommodating vessel. The expansion of the spring provides an estimation of the liquid density.

Gravitational Field

The area in which gravitational forces are highly functional. Its magnitude at any point because of the mass m is Gm/x^2 where x is the length of the point starting from m; its direction is across the line connecting the point to m.

Gravitational Waves

Waves disseminated as the outcome of the speeding up or distortion of a mass. The common theory of RELATIVITY anticipates the existence of gravitational radiation which travels at the speed of light. Coincidences between the signals from extensively distinct detectors, which comprises solid bars which must be set into oscillation through the waves, have been looked for. The consequences are claimed to signify those gravitational pulses from a point close to the midpoint of the Galaxy are arriving Earth; although, the evidence is debated.

Graviton

Refers to a hypothetical elementary substance, the quantum of gravitation, which is accountable for the impacts of gravity. It is anticipated to be its own antiparticle, to carry zero charge as well as zero rest mass along with a spin of 2.

Gravity Balance

Refers to a sensitive spring balance: the differences of the reading for a permanent mass at various places permits the differences of the acceleration because of the gravity to be analyzed.

Gravity Wave

A fluid surface wave having a large wavelength in which the motion is regulated by gravity instead of the SURFACE TENSION. Therefore, the immensity of the velocity v of shallow sea waves is provided by $v = (g\lambda)^{1/2}$, where g refers to the acceleration due to gravity and A stands for the wavelength.

Grease Spot Photometer

An optical photometer in which the photometer head comprises of a paper which is thin white opaque having a translucent spot at the middle. The two sources under contrast, A and B, of power PA and P8 consequently, are detected on either side of the spot so that a line linking them goes throughout normally across the center spot. The distances y_A and y_B of A and B consequently from the spot when it vanishes as perceived from one side are discovered, and also the correlating distances y'_A and y'_B for vanishing from the other side.

Grid Bias

A prospective variation applied between the cathode and regulation grid of a thermionic valve so that function performed on any chosen fraction of the characteristic· curve may be acquired.

Grid Leak

Between the cathode and grid of a thermionic valve, a high resistance is the grid leak. It ceases charge that gathers on the grid.

Ground Plane

Refers to a conducting sheet at earth prospective, implemented in an electric circuit in order to give a low-impedance way to earth all along the circuit.

Ground Wave

A radio wave transferred so that it moves through the surface of the Earth to the receiver, instead of reaching it through the reflection of *ionosphere.*

Group

A set of institutions having distinct mathematical attributes. Members of the set can be gathered through addition, multiplication etc. these institutions form a group if amalgamation of any two of them results into a third and if the mingling is an *associative operation*. In addition to this, there must be one institution, which is known as the identity, in a way that combination with it formulates no change. For instance, one is the identity connected with multiplication. Moreover, for each entity there must be another that amalgamates with it to capitulate the identity. A group member might be a number, or a collection of a *vector* or a *matrix*. Group theory is implemented in spectroscopy, crystallography, as well as particle physics.

Group Velocity

The velocity of dispersion of a pulse. The pulse may be regarded as resulting from the coinciding of sinusoidal waves of a bit distinct wavelength. For instance, with two such waves *beats* are acquired whose velocity of dispersion is $c - \lambda\ \delta c/\delta\lambda$ where c is the velocity (that is, the *phase velocity*) of one sinusoidal wave of wavelength λ and $c - \delta c$ is the velocity and $\lambda - \lambda\ \delta$ the wavelength of another sinusoidal wave. Usually, the group velocity refers to $c - \lambda\ dc/d\lambda$ and is the wave velocity as scientifically demonstrated.

Grove Cell

Refers to a two-fluid primary cell, the fluids being dilute sulfuric in addition to fuming nitric acids distinguished through a porous partition. The negative electrode is known as a zinc rod immersing into the sulfuric acid while the positive electrode is a platinum plate dipping in the nitric acid. 1.93 volt is the electromotive force.

Grown Junction

A semiconductor junction constructed by adding additional debasement to the melt whereas the crystal is growing in it.

Guard Ring

A big metal plate coplanar with and enclosed to a small metal plate yet distinct from it through an air gap. A implemented electric field is then stable across the small plate, which can be regarded as an infinite plane as the edge effects are withdrawn. An additional electrode which carries out an identical function is sometimes applied in semiconductor devices as well as vacuum tubes. A guard ring is also applied in accurate experiments on heat flow: the central plate together with the ring arc indcpcndently heated to the similar temperature so that edge impacts are removed and heat is fairly dispensed over the central plate.

Guard Wires

Earthed conductors positioned beneath overhead line conductors, so that, they will be earthed in the absence of striking the ground. Where high voltage lines goes beyond either telephone wires or a thoroughfare, a secured net of ground wires is applied.

Guillemin Effect

A variety of Magnetostrictions inside of which a magnetic field, implemented along the distance of a bent bar of ferromagnetic material, inclined to arrange the bar.

Gunn Effect

The formation of a consistent *microwave* current in specific homogeneous n-type semiconductors under the impact of an electric field of numerous thousand volt per centimeter. Such tools give microwave power sources that are in simple solid-state.

Gyrator

A device which alters the state of signals transferred in one direction, but carries no effect on the state of signals transferred in the opposite direction. It is predominantly applied at microwave frequencies.

Gyrocompass

A *gyroscope* which is left lose to swing in any plane, which revolves through an electric motor around a north-south axis. The axis therefore, sustains this direction regardless of any motions of the mounting. The device remains unaffected by stray magnetic fields and is extensively applied on ships, aircraft, missiles etc.

Figure 23. Gyrocompass

Source: Image by US National Archives

Gyromagnetic Effects

The associations between the magnetization of a body as well as its rotation. An iron cylinder which revolves around at high speed around its axis establishes weak magnetization and contrarily an iron cylinder which is free to rotate will do so moderately when randomly magnetized.

Gyromagnetic Ratio

The proportion of the magnetic dipole moment of a dipole at its angular momentum. Therefore, for an orbiting electron $\gamma = e/(2m)$, as well as for a spinning electron $\gamma = e/m$, whereby e and m refer to the electronic charge and mass consequently.

Gyroscope

Refers to a symmetrical wheel which is well-balanced and has its mass distributed in the farthest possible way from its axis and is ascended in a double gimbal so that it can spin without any hurdles about its axis, which can adapt any kind of orientation in space. The direction of the axis of the wheel repels transition, when the wheel is spinning. When put through to a considerably large torque inclined to change the direction of an axis, the gyroscope turns around a perpendicular axis to both its starting axis of spin as well as to the axis of the torque, that is the gyroscope exhibits Precession.

Gyrostabilizer

An instrument used on ships in order to stop excessive rolling. An immense and heavy Gyroscope might be applied to counteract the roll. On the other hand, a small gyroscope is implemented, this senses the roll and by means of a servo-mechanism activates fins that are stable below the level of water.

Gyrostat

It refers to a system of structures which contains relative motion that does not change the distribution of intrinsic mass of the system. For various given conditions gyrostats follows the Volterra equations of motions and for an inelastic body which rotates around a fixed point, it is a generalization of the Euler equations.

Figure 24. Gyrostat.

Source: Image by Wikimedia commons.

H

Hadron

Any substance that can take part in strong interchange. Hadrons come under two main groups, that is MESONS as well as BARYONS.

Hemorrhage Location

A technique most commonly achieved through inserting the patient along with blood categorized with the radioactive isotope that is chromium–51. This isotope is caught up through red cells and therefore, for blood circulation which is normal, the radioactivity is dispensed fairly all-round the circulatory system. An expansion of radioactivity at a few locations thus implies a hemorrhage at that specific site; the extent of build-up shows the amount of blood loss.

Hair Hygrometer

A HYGROMETER inside of which a hair is associated to a pointer is held tight through a spring. Expansion in the comparative humidity of the surrounding air is the foundation of the hair to expand in length, and to decrease in distance for a decline of relative humidity. Therefore, the pointer shifts, and it becomes easy to read off the relative humidity from a recalibrated scale.

Halation

The radiant patch enclosing the spot present on the screen of a cathode ray tube. It is most commonly is the result of the reflection that occurs internally inside the glass of the screen.

Half Cell

Refers to the one electrode of an electrolytic CELL in addition to the electrolyte through which it is in associated.

Half Life

The duration in which the rate of a radioactive nuclide decomposes to half of its real value. The half-life is equivalent to $(\ln 2)/\lambda$, where λ stands for the decay constant.

Half Period Zones

Zones throughout which a wave front is effectively and effortlessly split up when demonstrating the potency at a provided point which is a consequence of Fresnel Diffraction. The zones are of such type that the dispersion of light from one zone extending the point is half a period out of state with that reaching from adjoining zones.

Half Silvered Mirror

Refers to the MIRROR, the electroplated metal on whose edges surface is such that uniform proportion of light are passed on and reflected through it.

Half Thickness

The heaviness of a constant sheet of substance which when placed in the channel of a beam will transfer only half of it.

Half Wave Dipole

A direct aerial around half a wavelength long. When stimulated, the dipole has voltage antinode whereas current node at every end, while at the middle of it there exists a voltage node as well as the current antinode. The feeder is most commonly associated throughout a small gap in the dipole center.

Half Wave Plate

A plate of either quartz or mica cut collateral to the OPTICAL AXIS and of heaviness such that the usual and exceptional rays of transference light vary in state by 180°. Plane polarized light incident usually on the plate goes through an aspect of rotation of its plane of polarization, which is equivalent to twice the angle that exists between the axis and the incident vibrations' direction. The plates are applied in polarimeters.

Half Wave Rectification

The transformation of oscillating to direct current in such a manner that current moves through only in every other half cycles.

Half Width

It is a confusing and a complex term where either the width or half the width of a spectrum line is estimated at half its height.

Hall Coefficient

The coefficient RH in the articulation RHBJT for the prospective variation manufactured in the Hall Effect; B is the degree of the magnetic induction; J refers to the current density while T stands for the thickness of the specimen in the regulation of the prospective difference.

Hall Effect

The formation of a prospective variation between the surfaces that are opposite to each other of a conductor which is current-carrying, or even a semiconductor, positioned in a magnetic field which has direction perpendicular to both the differences that is of current and the potential variations. The prospective difference is the result of deflection of the carriers of charge and is in opposite directions for electrons and positive holes; its degree relies on the concentration of the charge carriers.

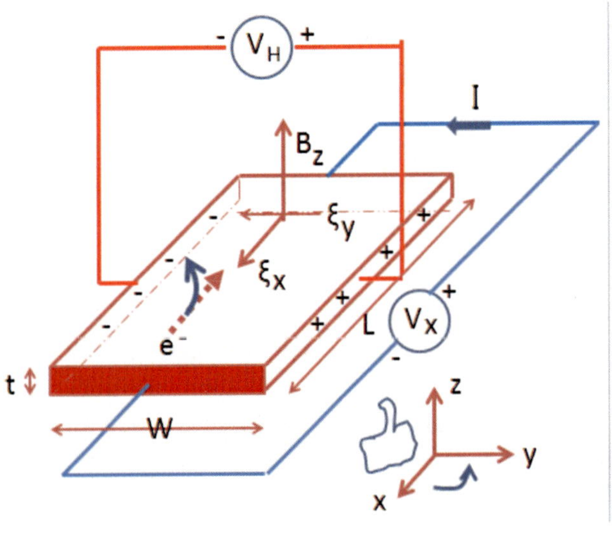

Figure 25. Hall effect.

Source: Image by Wikimedia commons.

Hall Probe

A small conductor or even a semiconductor of a well-known *hall coefficient* that is implemented to demonstrate magnetic installation by estimating the other variables in the light for the prospective variation.

Hardness

The opposition done by a crystal face to eliminating. It commonly differs with direction and might be estimated by the *Brinell test*. organized in order of decreasing hardness, some often applied materials are: diamond, corundum, quartz, topaz, felspar, fluorspar, apatite, calcspar, rock salt as well as talc.

Harmonic

A vibration that possesses frequency is an inherent number of times that of the basic, which is said to be the first harmonic. Compare to the *vibrations in pipes* and *vibrations in strings*. An uninterrupted function which assures the *Laplace equation* and which has its first derivative, a continual function.

Health Physics

The examination of health and security at work in medicine, science in addition to industry. Specifically, it deals with safety from *radioactivity*; this incorporates observation and estimation of the radioactivity, decontamination, discarding of radioactive waste, laboratory and shielding framework and the observing of tolerance of radioactivity doses accepted by workers.

Heart

The organ which sustains the movement and cycle of the blood through throbbing rhythmically at approximately 76 pulses per minute usually. The electrical pursuit of the heart, that denotes the spike prospective connected with the action of the heart, are accountable for opening and closing the valves of heart in the accurate sequence. An instrument which takes over the function and roles of the heart is called an artificial heart.

Heat

The energy transported from an extreme temperature body to a reduced temperature one as an outcome of the temperature variation between them. The heat of the body refers to the *kinetic energy* of translation, rotation and vibration in occurring molecules. For a body carrying out in the same phase, the transformation in its heat content because of the transformation in temperature is provided by mc9 where m and c are sequentially the mass as well as the *specific heat capacity* of the body. Transformation of phases of a body takes place at uniform temperature and is supposed to go along with the evolution or absorption of a degree ml of heat where 1 refers to the *specific latent heat*. Transmission of heat can take place through *conduction, convection* as well as *radiation*.

Heat Death

The phase of a closed system where if its average *entropy* has elevated to its highest value. Then, in that case, there is a lack of or completely no available energy since the temperature remains constant and every matter is disorganized. If the universe is an enclosed system, then it will in due course outreach to this state.

Heat Engine

A medium in which heat energy is transformed to mechanical energy. The engine captivates heat from a container at a given specific temperature and provided heat to a *heat sink* at a favorably decreased temperature; the cycle goes on repeating itself.

Heat Exchanger

A tool used for transporting heat in between fluids. The fluids might flow all along various channels distinguished by a wall, or flow either preferably or constantly by means of the same channel. An exchanger might be implemented for the purpose of cooling, as is evident in a car radiator, or for withdrawing the heat wasted from outlet gases, furnaces, or chemical plants etc. Heat exchangers may be given with devices or tools, for example, metal fins, coiled tubes etc. to the extended period of the time and region of contact of the fluids and so ameliorate the coherence and regulation of the exchange.

Heat Pump

A device which is implemented for the withdrawal of the heat from greater quantities of comparatively decreased temperature substances, such as air, and dispensing the heat at an extreme temperature. A volatile liquid evaporates in tubes in the presence of reduced temperature material, the required heat coming out from that substance. The vapor is then compacted and reliquefies, giving rise to the heat.

Heat Sink

A device used for managing and scattering heat away from electronic elements so as to cease the destruction towards them. It is a system regarded in order to absorb heat at a fixed temperature, as in the regulation of a heat engine.

Heat Transfer

A method which may take place in three manners that are *conduction; convection* and *radiation* respectively.

Heat Transfer Coefficient.

The degree of heat transmitted per second per unit area through a surface at a given temperature of I K exceeding its surroundings.

Heavy Atom Technique

A technique used in the examination of molecular framework by X ray diffraction. It incorporates the introduction into the material that are under observation of atoms which radiate X rays well, that is atoms having high atomic number, and mathematical treatment of the acquired outcomes; by consecutive estimation a considerably precise structure is reached at.

Heavy Water

A liquid having the identical chemical attributes as water but with distinct physical properties. The term is precisely applied for deuterium oxide, which carries the hydrogen isotope deuterium along with a comparative atomic mass of 2. Heavy water is released from natural water by two methods, one is either by fractional distillation technique or by electrolysis or in other words exchange processes.

Heisenberg Picture

The demonstration of dynamic phases of a system which is quantum mechanical by stationary vectors, and of physical amount by engineers which keeps on growing with time.

Heisenberg Uncertainty Principle

The principle that it is not possible to estimate both the momentum and location of a particle to any desired amount of perfection. Errors in estimating instruments and errors made by human that are not accountable for this indefiniteness, which grows from the behavior of particles which is wavelike and from the examination itself disrupting the system in an unanticipated manner.

Helium

Helium has a Symbol which is He. A substance available in some natural gas, in few of the radioactive ores in addition to the atmosphere, from which it is withdrawn in the form of a byproduct of the liquefaction of air. It is applied for stuffing air ships and several kinds of fluorescent lamps. Liquid helium lives in two apparent forms, I and II, sequentially uniform above and below 2.19 K. Form II demonstrates superfluidity and contains high thermal opacity as well as low viscosity.

Helix

A curve which is drawn on thc surface of a cylindrical or conical in a manner that all the inductions of the surface are divided at the given identical angle.

Heterodyne

Using or involving the radio frequency version of beats. If the radio signals are close in frequency the heterodyne frequency difference is in the audio – frequency range, an effect used in radio reception.

Heterodyne

It is a radio reception effect that uses a radio frequency version of Beats. In this the heterodyne frequency is considered in the audio frequency range when the frequency of radio signals is close.

Heterogenous Radiation

It is a an infrared or visible non monochromatic radiation

Heterogenous Reactor

The core of a heterogeneous reactor which is a nuclear reactor includes moderator embedded thin fuel rods.

Higgs Particle

Electroweak theory predicted the existence of Higgs particle and association of z particle and w particle are believed to be a part of its massiveness.

High Pressure Technique

High pressure technique is required to generate pressure of 1 gigapascal to 6 gigapascal with the help of crossing piston into cylinder. High pressure technique is associated with research on geological phenomena, phase equilibrium work and compressibility studies.

High Tension

High tension ranges from 60 volt to 250 volt and this voltage is applied to thermionic valve's anode

High Voltage

For electrical power transmission and distribution High voltage is considered which is more than 650 volts

Hilbert Space

It is a multi-dimensional space in which representation of every characteristic function of wave mechanics is considered by an orthogonal unit vector

Histogram

For graphical representation of data, the range of values is plotted along the x-axis divided into equal intervals in which the base of a rectangle is represented by every interval and the y co-ordinate represents height of the rectangle.

Hole

For the electronic energy level in a solid the unfilled vacancies are called holes that are mathematically represented by positrons. Holes are the majority carriers in a p-type semiconductor and electric conductivity is dependent on holes.

Hole Conduction

Under the influence of an electric field holes move through a solid when the hole is filled by an electron and it creates a hole in the position from where the electron came from. Similarly, more electrons fill the holes and this movement is similar to movement of positive charge.

Holography

In this technique lenses are not used for obtaining a stereoscopic image of an object. Using a half-silvered mirror, the laser beam is divided into two parts the object is illuminated by the first part and with the help of second part is used to reflect or transmit the object over a high-resolution photographic plate. The information of phase and intensity is carried by the image on the plate. With the help of a microscope a large number of closely spaced dots can be seen with the help of this photograph which are called holograms. Diffracted waves are produced with the help of direct elimination of the hologram that is produced by the same laser light. The phase distribution and amplitude are similar to the original ways from the object.

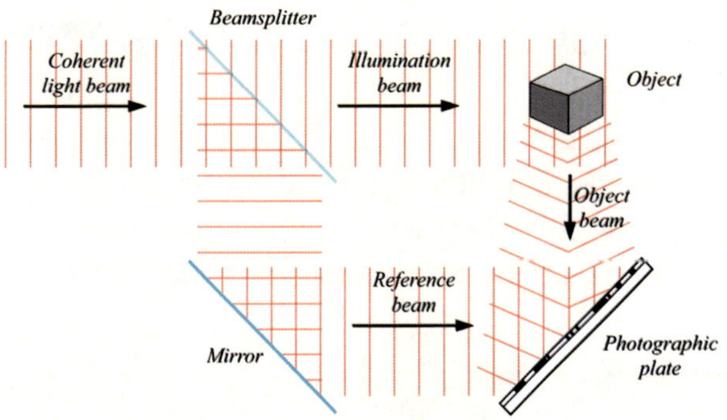

Figure 26. Holography.

Source: Image by Wikimedia commons.

Homopolar Generator

Small direct voltages are generated by such polar generators required in electroplating. The copper disc used in this generator has brush connection to the circumference and Excel and rotates between the poles of a magnet at r revolution per second

Hot Wire Ammeter

By increasing the temperature, the wire is expanded so that current can be measured that passes through a wire. Over a graduated scale the rotation of a pointer is produced with the expansion. As the heating effect is directly in relation to the square of the current then both direct current or alternating current can be measured. By implementing series resistance this ammeter can be used as a voltmeter.

Hot Wire Anemometer

With the help of fluid movement over the heated wire the cooling effect that is produced is used to measure the fluid velocity by this instrument. Wheatstone bridge is used to measure the change in resistance that is required to measure the change in temperature of the wire. Usually, a galvanometer is used with the bridge stone bridge but in case when the fluid velocity where is then the cathode ray oscillograph detector is used.

Hot Wire Gauge

The cooling process of the hot filament by the gas depends on a gas pressure.

Hot Wire Microphone

In order to measure the amplitude and intensity of sound waves a device is used that uses electrically heated fine wire and resistance decreases in the wires caused by impact between the wire and sound waves.

Hubble Constant

The ratio measured by redshift between galaxy's recessional velocity and its distance. It's not generally considered to be a true constant and the value ranges between 1.6×10^{-18} per second and 3.2×10^{-18} per second. The universe expansion rate is provided to be constant and the age of the universe is given by 1/H whereas its upper limit is 20×10^{9} years. Additionally, c/H, c is the speed of light and this provides radius of the observable universe the value has an upper limit which is 1.85×10^{26} meter.

Hum

The amplifier output provides an extraneous alternating current called HUM which originates in a nearby electric power circuit.

Humidity

- Absolute Humidity. It is defined by the presence of mass of water vapor in unit volume of air. The value is 0.002 17e/T, where 'T' is the thermodynamics temperature and 'e' denotes the water vapor pressure
- Relative Humidity. It is represented by u when water vapor is used as a saturation agent for Humidity in air and humidity at the same temperature then the ratio of actual absolute humidity and absolute humidity of air is called relative humidity. This is represented by percentage.
- Specific Humidity. It is the water vapor mass present in air per unit mass of the moist air.

Hybridization

Equivalent and directed orbitals of an atom are combined to form hybrids and this process is called hybridization.

Hydrated Electron

In the presence of a collection of water molecules when an electron is found to be at the center of this collection is called a hydrated electron. These electrons are very reactive in nature and they occur when aqueous solution is irradiated with ionizing radiation.

Hydraulic Press

It is a pressure applying device. It is commonly known that the liquid compression process is very difficult to follow and hence the pressure applied to the liquid surface is transmitted through it and distributed over a large area. Hydraulic pressure is used in this case.

Hydroelectric Power Station

Process of conversion of hydraulic energy to electrical energy is performed in a power station called hydroelectric power station. In order to drive a generator, energy provided by a water turbine is used in which kinetic energy is acquired by the water falling from top level to bottom level.

Hydrogen Electrode

It is used for measurement of electrode potentials. In a solution of hydrogen ions hydrogen is applied in an electrode system. Half-cell is used in this system where hydrogen gas is bubbled over platinum foil which is immersed in a dilute acid.

Hydrometer

The relative density of a liquid is used to measure by an instrument called hydrometer. With the help of a length of calibrated stem that is emerged from the liquid the relative density of the liquid is determined. The long-calibrated stem is always vertically floating when it is immersed in liquid in a constant mass hydrometer.

Hydrostatic Balance

Relative density of a substance is measured with the help of beam balance. The substance whose relative density has to be calculated is suspended in such a way that it is balanced above the pan; a wooden bridge is used to provide the balance that is rested across the pan but not touching it. In order to calculate the relative density, the substance is weighed toys that is first it is weighed in air and then in water. With the help of these observations the relative density of the substance is observed.

Hygrometer

To measure the relative humidity of air an instrument called hygrometer is used.

Hygroscope

It is a device that is used to indicate the approximate value of relative humidity of air. In this device Cobalt Chloride is used that is impregnated by a card so when the change in color is seen changing from Blue to Pink then it is observed that atmosphere is relatively humid because its moisture content is high.

Hyperbola

Eccentricity of a cone when measured to be more than one then it is called hyperbola. It is located on the locus of a point and the difference between distance from the foci is always constant. Where is the color rectangular cartesian coordinates equation $x^2/a^2 - y^2/b = 1$, in this equation a and b are constants and the asymptotes are the curve approaches.

Hyperboloid

Consider the equation of hyperbola as $x^2/a^2 - y^2/b^2 = 1$, which is rotated about either the x axis or the y axis and a solid of revolution of tenth by this process is called hyperboloid. To sheet form is yielded when the rotation is about x axis and one sheet form is obtained when the rotation is about y axis. Solids in which cross section is not circular and it is electrical is also covered by hyperboloid.

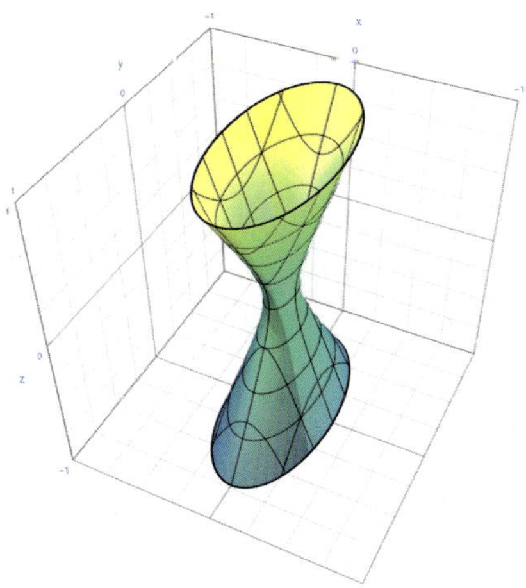

Figure 27. Hyperboloid.

Source: Image by Wikimedia commons.

Hyperfine Structure

In order to obtain monochromatic radiation at low resolution apparatus called Fabry-Perot interferometer is used in which 10- to -12-meter range of wavelengths appears.

Hypermetropia

It is an eye defect in which accommodating power of the eye is used to see distant object while for distinct vision of near objects a convex lens is used because parallel rays of light are focused behind the retina.

Hyperon

Hyperon is not a nucleon but it can be considered to be a Baryon because it participates in strong interaction and has a mass which is more than proton and it also has half integral spin. Examples of hyperon particles are Lambda, sigma, xi, and omega minus particles. Nucleons are formed after hyperon decay.

I

Ice Point

The standard atmospheric pressure at which water and ice are in equilibrium and the temperature is called ice point. Previously on the international temperature scale it was defined as 0 degree Celsius but now fixed point by the triple point of water has replaced it which is more than the ice point.

Icosahedron

It is a 20 faced polyhedron. The icosahedron is called regular when these are congruent equilateral triangles.

Ideal Crystal

There are no impurities and defects found in idle crystals and the lattice of ideal crystal have perfectly regular structure.

Ideal Gas

Atoms of ideal gas have perfectly elastic collisions and these are composed of negligible volume atoms. It is confirmed By Dalton's Law of Partial Pressures, Joule's Law, Charles' Law, Boyle's Law and Avogadro's Hypothesis. At low pressure real gases act like ideal gases.

Illuminance

The energy of light striking a surface per unit area per unit time, i.e., $d\phi_v/dA$ where ϕ_v is the Luminous Flux and A the area.

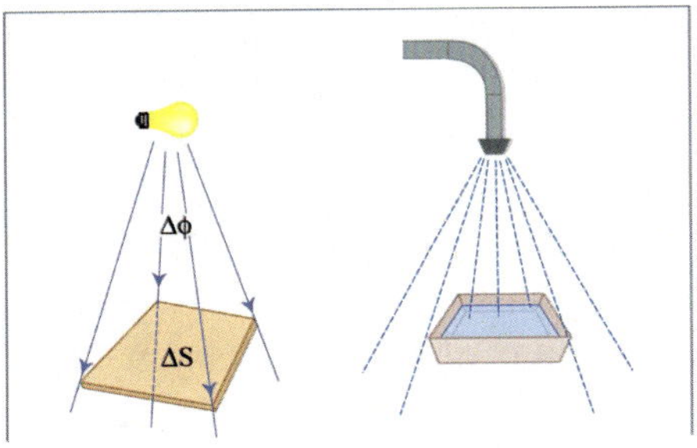

Figure 28. Illuminance.

Source: Image by Flickr.

Image

When an object is represented on a surface it is called its image. Real images are formed when light rays pass through the object and these are convergent if the light rays do not pass through the object but diverge from the image then the image of the object is called virtual image. Difference between a real image and virtual image is that a virtual image cannot be represented on a screen while a real image can. With the help of physical optics and geometric optics the study of image formation by light can be done.

Image Converter

The infrared type invisible radiation is used to obtain a visible image with the help of an electronic device called image converter. Electrons are released with the help of a photocathode on which radiation is focused. Visible images are produced when fluorescent screen anode that is positively charged attracts the electrons.

Imaginary Number

Negative number' square root is called the imaginary number. This number is denoted by ia, where i is the square root of -1 and a is a real number.

Immersion Objective

Optical microscope requires an immersion objective. The microscope specimen slide has a cover glass on top of which is a liquid placed which emerges to the lens nearest object. The liquid used is either sugar solution or cedarwood oil that has a refractive index similar to that of a cover glass of microscope. The resolution of the microscope and the numerical aperture of the objective are increased with the help of liquid.

Impedance

- Electric Impedance. Electric impedance is represented by Z. To resist the current flow of an AC circuit is measured by a complex quantity. The ratio of root means square voltage to root mean square current is the magnitude of electric impedance.
- Acoustic Impedance. Acoustic impedance is the sound passage resistance of a sound wave carrying device.

Impedance Matching

Circuits are connected devices for which suitable impedance values are selected; this process is called impedance matching. When power is transferred from one device to the other then it becomes very important to consider the impedance matching procedure to gain maximum efficiency in power transfer and it should not be just left to be set on its own.

Impedance Plethysmography

Conducting regions of the body have volume changes that are detected with the help of a technique called impedance plethysmography in which changes in electrical impedance is measured for measuring volume changes. The eternal blood flow and breathing are monitored with the help of this technique. In order to measure blood flow legs or arms are attached with leads while for breathing measurement it is attached to the chest. In order to measure the impedance, change low voltage frequency range is applied to the pair of electrodes. It is important that current flow should be as low as it should not be noticed by the patient in that case another electrode pair amplifies the signal to observe the measurements.

Imperial Units

Imperial units are the measurement units in which lard is used for unit of length and pound is used for unit of mass and this system is getting replaced by SI units.

Impurity Semiconductor

It is an extrinsic semiconductor that includes either externally introduced or naturally occurring foreign atoms that impacts the type and amount of conductivity and overrides this effect. Impurity semiconductors are used to increase the conductivity of the semiconductor.

Incandescence

When the wavelength of a radiation emitting substance lies in the visible region of the spectrum it is called incandescence.

Induced Electromotive Force

Due to suitable movement of the conductor or by actual change in the flux the magnetic flux linked with the conductor changes and it produces an electromotive force which is called induced electromotive force.

Inductance

When induced electromotive force is applied in opposite direction of the current flow then resistance to the flow of changing current is measured by the electric circuit which is called inductance.

Induction Coil

It is a device which with the help of electromagnetic induction produces high voltage. In the induction coil there are two coils, one is the primary coil with few turns of insulated wire and the other is the secondary coil that has many turns of insulated wires on a laminated iron core. The secondary coil surrounds the primary coil. Primary coil is fed with a low current that is interrupted by a contact breaker in secondary coil in order to produce voltage output. When direct current is required to produce intermittent high voltage then an induction coil is used for this process of High voltage generation from direct current. According to the induction lock value of windows electromotive 4 is directly dependent on the magnetic field to the coil as its time rate of change.

Figure 29. Induction coil.

Source: Image by Wikimedia commons.

Induction Heating

With the help of alternating magnetic field Eddy currents are induced on metals to heat them this is called induction heating.

Induction Motor

It is a commonly used AC electric motor in which the rotating magnetic field of the stator winding is used to produce electromagnetic induction that allows producing electric current in the rotor. The rotor can be wound type rotor or squirrel cage rotor used in an induction motor.

Inequality

When Quantity is required to be shown not equal then mathematical statement of inequality is used. According to this is a>b, a is greater than b that means that a and b are not equal similarly in this case a<b, the inequality statement shows that a is less than b.

Inertial Force

Inertial force is acted in the opposite direction on a body that is already being under the effect of an accelerating force. The inertial force is measured by the product of mass of the body and accelerating force exerted on the body.

Information Theory

In communication when a specified problem is required to be solved then the minimum amount of information required to solve the problem with the help of an analytical technique called information theory.

Infrared Astronomy

With the help of infrared radiations astronomical objects are observed in infrared astronomy. In the range of 0.75 micrometers to 300 micrometers lies the wavelength of infrared light. While the range between 380 nano meters to 750 nano meter is for infrared to fall in the visible radiation.

Infrared Radiation

In the electromagnetic radiation spectrum, the range of wavelength from 700 NM to 1 mm is the reason in which infrared radiation falls. Infrared waves are in between visible light and radio waves as it is shorter than radio waves but longer than visible light. Also infrared frequencies are considered to be higher than microwaves but its frequency is lower than that of visible light, that is the range of its frequency lies in between 300 Ghz to 400 THz. Human eye is not able to see infrared light but it can be changed as heat when longer infrared waves are sent. The characteristics of infrared light and visible light that are similar are that both of these lights can be polarized, reflected or focused.

Infrared Spectrophotometer

Materials like organic polymers are identified with the help of an analytical instrument called infrared spectrophotometer. When the infrared radiation is made to pass through a sample then this analytical instrument considers wavelength function as a relative amount of energy to record it.

Infrared Spectrum

The absorption spectrum that falls in the infrared region is called the infrared spectrum. The molecular vibrational frequencies are determined with the help of infrared spectrum and it is subjected to study for this purpose.

Infrasonic

The lowest audible frequency is 20 Hz and the frequency of sound below this frequency is called infrasonic.

Injector

In order to mix two fluids together a device called an injector is used in which the principle of fluid mixture states that when in a constricted region fluid is pressurized to pass in a pipe then this pressure is less than the pressure in the part of pipe which is unconstructed. It can also be understood from the point that fluids of low pressure are made to enter a higher-pressure region with the help of a device called an injector. To control the inlet, flow a valve is used otherwise there are no moving parts in this fluid dynamic pump.

Input

Any information that is fed into a system to pass through a predefined process so that it can provide an output is called input. For example, an electric circuit is fed with signal, current or voltage as input. Similarly, the data fed to the computer is also input.

Integral Equation

It is a mathematical equation in which an integral sign is used to show an unknown function in the form of an equation. In order for an integral equation to be linear the unknown function needs to be linear as well and similarly a function needs to be determined in order to form a solution of the equation.

Integrated Circuit

It is a set of electronic circuits that is integrated on a small silicon piece which is a semiconductor material.

Integration

Integration is a process of determining integrals which is used to describe the infinitesimal data that is provided to a function of volume, area or displacement.

Interference

Wave trains of the same frequency and same phase are interacted then it produces progressive waves that is called interference. Audio frequency amplifier or simply called amplifier is used to generate the sound waves with the help of parallelly connected loudspeakers. For example, interference can be observed by an observer when they are moving in a straight line and get to experience two different zones one is silent and other is loud. This happens for silent zones when the waves cancel out each other because they are out of phase as the path difference is an odd number between the observer and the sound waves. Similarly in the loud reason the sound waves are in phase and whole number determines the he wavelength for path difference and hence the sound waves reinforce each other from two different parallelly connected loudspeakers.

Interferometer

An instrument called Fabry-Perot Interferometer is designed so that it can generate interference fringes. This apparatus is used to determine the surface flatness, small distances and wavelengths.

Intermolecular Energy

Intermolecular energy is of two types that contains thermal and other is in the form $a/r^p-b/r^q$. The first one is dependent on the molecules kinetic energy that can be potential or temperature caused when molecules interact electrically.

Intermolecular Force

Combined force of two neighboring molecules is called intermolecular force. This force is generated from the kinetic energy of the molecule and the electrical charges that the molecule has on its different parts that may intervene the neighbors. Intermolecular forces are categorized in three parts that are ion dipole interaction, dipole–dipole interaction and London dispersion forces. Hydrogen bonding also contributes to the intermolecular force because it is a dipole – dipole interaction form. Simple it can be defined as the forces exerted between two neighboring molecules in an atom.

Internal Conversion

When for an excited state the Alpha radioactive decay leaves the nucleus in the case of a gamma emission, it causes nucleus de-excitation and is known as internal conversion. The energy is transferred to the electron of an atom when gamma emission or internal emission interacts with this electron; this process can also be called electronic conversion. The electrons are bound to the atom by some forces and it should escape these forces in order to inherit the gamma energy.

Internal Energy

A substance contains microscopic energy e in it that is provided in the form of kinetic energy of the molecules that are this order and hence these are called internal energy. Between two molecules there is some potential energy which is included in the internal energy and in the atoms of the molecules there is some nuclear energy which is also part of internal energy. It is represented by U and measured by combining all the potential energy and kinetic energy of the atoms. State of the system is responsible for determining internal energy of the system. The change in internal energy can be measured but its absolute value is not known. Systems initial and final states are responsible for internal energy changes and not how the states changed.

Internal Resistance

Hindrance in the flow of current that is responsible for heat generation is called internal resistance. It is measured in ohms. It is represented by (E–V)/I; where E is the electromotive force and V the potential difference across the source when supplying a current I.

Internal Work

There are some intermolecular forces responsible for holding the atoms together and when the molecules are required to be separated then the work done for the separation is called internal work. The internal work is always zero in the case of an ideal gas.

International Temperature Scale

The combination of temperature measurement unity techniques, interpolation dependencies and reference point numerical values are called international temperature scale.

Interpolation

The process of calculation of the function value of a variable that ranges between values in which function value is known.

Interval

All real numbers that lie between two numbers of the set are called an interval for a set of real numbers. In the essence of time interval can be determined by the equal set of time periods called time intervals that can be denoted by second, minutes or hours.

Intrinsic Semiconductor

These are pure semiconductors because there is no impurity found in it and it also has no real existence but for nearly pure materials the term is used. When the hole densities and electrons in a semiconductor are equal then it is called an intrinsic semiconductor.

Invar

It is an alloy with lower thermal expansion coefficient. It is shorter form of invariable and it is given this name because with change in temperature there is lack of contraction or expansion in the alloy hence it is given the name in variable or invar.

Inverse Square Law

Inverse square law represents that when the physical properties like force get separated farther then they get smaller. So, when interacting objects get separated more and more than that property's intensity decreases. Examples are Coulomb's Law Lamberts Law and Newton's Law of Gravitation.

Inversion

- A mathematical transformation.
- It is a process going in the usual direction when reversed is called inversion for example heating of water and its behavior change in this process.
- An optically active substance is transformed into a rotatory effect that is opposite to the previous one without intervening in the chemical composition of the substance.
- A form of speech scrambling.
- In the semiconductor surface when an opposite type layer is produced then the process is called inversion.

Inversion Symmetry

Under the effect of mathematical operation of inversion, the physical laws are said to be invariant and this theory is called inversion symmetry but the theory is violated by weak interaction.

Inversion Temperature

In cryogenics and thermodynamics inversion temperature is determined as that critical temperature above that temperature a non-ideal gas will experience a temperature increase as it was expanding at constant enthalpy and below the critical temperature it will experience a temperature decrease.

Iodopsin

A protein which contains a retinene group located in the receptors of human retinal. In the current times, its function is not understood because of the fact that its absorption spectrum does not match any visual response curve.

Ion

An atom or group of atoms having a net charge. Positive ions contain a shortage of electrons while the negative ions have an excess of electrons.

Ion Implantation

The shelling of a semiconductor together with ions of adequately high energy to creep into the surface. Regulated doping of the semiconductor might therefore be accomplished.

Ionization

The method of constructing ions either from neutral atoms or molecules. Ionization might take place randomly when a substance is disintegrated. Other techniques of the construction of ion incorporate captivating electrons, bombardment with ionizing radiation, photon bombardment in addition to thermal ionization.

Ionization Chamber

A device implemented for examining ionizing radiation. It comprises two electrodes which contains opposite charge arranged in a chamber having a gas. The gas atoms or molecules are ionized along with this the ions moves along to the electrode of opposite sign, when ionizing radiation penetrates through the chamber. This results in an ionization current which, under appropriate circumstances, is in proportion to the intensity of the radiation.

Ionization Gauge

A vacuum pressure gauge having a three-electrode thermionic tube blended to the gas system which contains pressure that is to be estimated. Electrons which move along the grid ionize gas molecules positively so that they pass through the negatively charged plate. This formulates a plate current which is an evaluation of the number of gas molecules available. Pressures can be measured which is as low as 10^{-3} pascal.

Ionization Potential

The minimum energy needed in order to withdraw the least strongly bound electron to a boundless distance from its parent atom and so generate a positive ion. The least energy which is needed to withdraw the second least strongly bound electron out of a neutral atom or molecule is known as the second ionization potential and so forth.

Ionizing Radiation

A flow of particles or a beam of electromagnetic radiation which results in ionization in moving along a medium. For this to take place, the particles' photon energy of the radiation or the kinetic energy must go beyond the ionization potential of the substance under pelting. Gamma rays, X rays, electrons, protons and alpha particles are the classic examples of ionizing radiation.

Ionosphere

An area of ionized air as well as free electrons expanding from a height of 50 kilometers to 1000 kilometers exceeding the surface of the Earth. It is the result of X ray together with the short-wave ultraviolet ionizing radiation from the Sun. considering that the free electrons present in the ionosphere reflect radio waves, in this case, radio transmission becomes possible between points concealed from each other by the curvature of Earth.

Ion Pump

A pump implemented for the construction of a very high vacuum. The remaining of gas is ionized through an electron beam and the resulting positive ions are gathered at a cathode where they left their confined. The pump is only beneficial at a pressure which is less than 1 micro pascal. To tackle with the difficulties of cathode saturation through ions, the sputter ion pump has been developed. In this instrument, electrodes of titanium, which are spattered by a release, are made use and therefore a film of titanium which performs as a getier is constantly constructed.

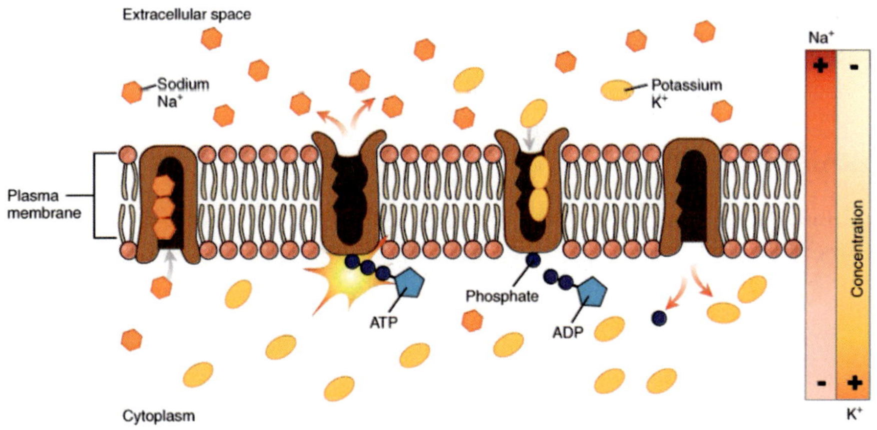

Figure 30. Ion pump.

Source: Image by Wikimedia commons.

Irreversible Process

A method in a way that, when concluded, the system incorporates cannot be revived to its initial thermodynamic state. All processes that are natural physical are permanent.

Isobar

Refers to a line on a weather map which moves along regions of equal atmospheric pressure. Any of a number of nuclides containing the same mass number but having distinct atomic number.

Isomorphous Replacement Technique

A process implemented in the diffraction of X ray analysis in order to construct the structure of complex molecules for which the heavy atom method is of small use. The technique comprises in forming two compounds similar apart from the replacement of a specific heavy atom by another; by examining the X ray patterns from the two compounds it is plausible to reduce the position of the exchangeable atoms.

Isothermal Bulk Modulus

For constant temperature conditions the bulk modulus of elasticity is known as isothermal bulk modulus.

Isothermal Process

A process taking place at constant temperature.

Isotones

Nuclides with the identical neutron number but having dissimilar atomic numbers. For instance, $^{39}_{19}K$ and $^{40}_{20}Ca$ each carries 20 neutrons.

Isotropic

Containing properties which are independent of direction. Therefore, an isotropic crystal carries the identical physical properties throughout all its axes.

J

Jaeger's Method

A practical technique of evaluating surface tension of a liquid at several different temperatures. An air bubble is constructed inside the liquid on the end of a glass Capillary tube having the radius r. The surplus pressure inside the bubble is estimated and the surface tension y demonstrated from the formula $y = rg (hp - h_1p_1)/2$, h1 is the depth of the end of the capillary lower to the surface of the liquid which has the density p_1 at temperature t_1; h is the pressure gauge reading, where p the density of the pressure gauge liquid while g stands for the acceleration due to gravity. By carrying out the experiment again and again for different values of t_1, the differences of surface tension along with temperature for the liquid can be demonstrated.

Jamming

The intervention of the radio reception in a way that the desired signal is incomprehensible.

Jet Propulsion

Forward propulsion which is generated by one or more jets having high-velocity hot gas supplying from conventionally regulated nozzles.

Jet Tone

The variable hiss going along with the appearance of a stream of mobile air into uniform air.

Joly's Steam Calorimeter

An apparatus which is applied evaluating the specific heat capacity of a gas at a given fixed volume. Two copper spheres, as closely similar as possible, are suspended from the opposite arms of a beam balance.

The spheres are enclosed by an enclosure by the medium of which steam can be progressed. One sphere is evacuated while the other suffused with a mass m of the sample gas at an extreme pressure. When both spheres go to the steam temperature t5, the surplus mass of water m5 precipitated on the gas-filled sphere is documented.

Then $C_v = m_s l / m (t_s - t_r)$ where L stands for the specific latent heat of steam and tr refers to the room temperature. To compensate for any variations in the spheres, a repeat regulated experiment is carried out with both spheres evacuated; m. is rectified by any variation in the masses of steam precipitating on the two spheres. The spheres are shielded from drops of water falling from the compounded roof. Any drops falling from the spheres are captivated and so incorporated in the estimation of the mass.

Josephson Effect

The taking place in two superconductors, distinguished by a thin dielectric, of an oscillatory current of frequency which is in proportion to a steady prospective difference implemented between the conductors.

Joule

The symbol for Joule is J and it is the SI unit of energy, which is equivalent to the work done when the point of application of a force of volume one newton is progressed one meter throughout the line of action of the force. The joule also stands for the work done through a current of one ampere flowing along a resistance of one ohm for one second.

Joule Heating

The quantity of heat generated in a conductor through an electric current flowing along it. If a stable current of I ampere flows fort second in the medium of a resistance of R ohm, the heat generated is I^2Rt joule.

Joule-Kelvin Coefficient

An amount which is connected with the Joule-Kelvin effect and is equivalent to the rate of transition of temperature together with pressure at persistent enthalpy.

Joule-Kelvin Effect

The transition in temperature of a gas when experiencing a technique of adiabatic evolution. It may be demonstrated experimentally by enabling a volume V of air at room temperature as well as high pressure p to pass through a porous plug throughout a vacuum; the air temperature is examined to fall by the difference of around 0.1 K per atmospheric pressure.

The enthalpy of a gas is permanent. Thus, at initial gas temperature if pV expands (decreases) asp reduced (increases), then in that case, the internal energy U reduces (increases) with along with p. A reduction in U is connected with a fall, while an increase in U with an increase in gas temperature. Variance from Boyle's Law may therefore construct either heating or cooling.

In comparison, deviation from Joule's Law formulates only cooling due to the fact that the potential energy of the attractive forces present between the gas molecules is expanded either by increased volume or by decreased pressure.

J/PSI Particle

A Meson of mass 3 gigaelectron, which consists of a charmed Quark as well as a charmed antiquark. Its revelation provided support to electroweak theory in addition to the charm concept, and impulsion to the evolution of the theory of strong interaction.

Jupiter

A giant planet consisting of diameter 142,800 kilometers. Its mass is 318 times that of the Earth whereas its density is only one fifth of Earth's. Its gravity is 2.5 times as well as its magnetic field is approximately 17,000 times those of Earth. Jupiter is 778 million kilometers away from the Sun and is considered the fifth planet out from it.

Figure 31. Jupiter.

Source: Image by Flickr.

Its orbital together with the average axial rotation periods are 11.86 years and 9-hours 51 minutes sequentially. The atmosphere of Jupiter comprises of hydrogen, helium, methane, nitrogen and ammonia. The southern hemisphere is identified by what is called as the Great Red Spot. The subject of examination and studies by US space probes have always been about the planet and its various satellites.

Juvin's Rule

The formula $h = 2\gamma \cos \alpha /(rgp)$ where 'h' stands for the variation in height between the level of liquid inside as well as outside a tube which is vertical open-ended capillary tube, internal radius r, positioned in the liquid; p and γ are sequentially the density and surface tension of the liquid, g is the acceleration because of the gravity and α refers to the angle of contact between the surface of liquid and the wall of the capillary tube.

For liquids which moisturizes the tube, α is less than 90° and so because α is positive; Thus, these liquids elevate in a capillary tube. For liquids which do not moist the tube, α is larger than 90° and so because α is negative; Thus, such liquids fall outside the tube which occurs below the general liquid level.

K

Keeper

A small iron or steel component needed for completing a magnetic circuit and avoiding magnetism loss when magnets are in storage.

Kellner Eyepiece

A Ramsden eyepiece variant with a cemented eye lens that rectifies for chromatic *aberration* and distortion more effectively than the original. Prism binoculars mostly employ it.

Kelvin

Symbol K. The SI unit of thermodynamic temperature. It is given as 1/273.16 of the triple point of water's thermodynamic temperature. One Kelvin is equivalent to one degree Celsius on the temperature scale, i.e., 1K=l 0 C.

Kelvin Contacts

A method to evaluate electronic circuits and components that negates the effect of lead resistance on readings.

Kelvin Double Bridge

A Wheatstone bridge that is employed for reliable measurement of a low resistance R. R' has the same order of resistance as R. Throughout, the relation $R_1/R_2 = r_1 / r_2$ is constant, but its value is modified until the galvanometer G shows zero current. The resulting ratio is R/R.' Errors caused by contact and lead resistances are eliminated.

Kelvin's Formula

In an electric circuit with minimal resistance, it is given by the formula $T = 2\pi(LC)^{1/2}$, where T denotes the period, L denotes the inductance, and C denotes the capacitance.

Kepler Telescope

A refracting astronomical telescope with convex lens systems used for the objective and eyepiece. The focal points of the two systems coincide when utilized in normal adjustment. An erecting prism is fitted into the device to modify the instrument to a terrestrial telescope.

Figure 32. Kepler telescope.

Source: Image by Wikimedia commons.

Kerma

The ratio of the sum of the initial kinetic energy of charged particles created indirectly by ionizing radiation in a tiny quantity of substance to the mass itself.

Kerr Effect

An electro-optical effect which takes places when some liquids and gases are exposed to a strong electric field perpendicular to the direction of illumination, inducing double refraction. If n_1 and n_2 are the refractive indices for ordinary and extraordinary rays, respectively, then $n_1 - n_2 = k\lambda E^2$, where λ is the wavelength of the radiation employed, E is the magnitude of the electric field strength, and k is Kerr's constant.

The phenomenon has a practical use in the Kerr cell, a transparent device with a suitable liquid and two electrodes, which functions as a shutter: by providing an appropriate voltage across the electrodes, a beam of plane polarized light incident on a Kerr cell can be stopped. A magneto-optical effect wherein a plane-polarized light beam striking the polished pole face of an electromagnet reflects with little elliptical polarization.

Kew magnetometer

A magnetometer that uses a steel tube as a magnetic needle. A convex lens is attached to one end of the tube, and a numbered clear scale located in the focal plane of the lens is attached to the other. In normal adjustment, the scale is examined through a telescope. The device is used to precisely measure the Earth's magnetic field.

Kinetic Energy

Symbol T. The energy preserved in a system as a result of mass movement within the system, as measured by the work required to return the system to rest. The kinetic energy of a body with mass m and speed v is equal to $mv^2/2$. The kinetic energy of a body whose moment of inertia and angular speed about an axis of rotation are I and w, respectively, equals $Iw^2/2$.

Kinetics

- The study of the rates of chemical processes. It provides detailed information about reaction mechanisms.
- The dynamics of physical bodies.

Kinetic Theory of Matter

A theory that connects the motion of individual molecules and a substance's macroscopic features. Intermolecular forces are so powerful in solids that molecular motion is mostly limited to vibration around a fixed point. Dulong and Petits law results from the use of kinetic theory. As attractive forces between molecules are weaker in liquids than in solids, molecules move around at random primarily inside the liquid.

Klein-Gordon Equation

In nuclear quantum theory, a relativistic variant of the Schrodinger equation is used: $\nabla^2\psi + [(E\text{-}V)^2 - m^2c^4]\ \psi/\ (h^2c^2) = 4\ \pi\rho$, where ψ is the Schrodinger wave function, E is the total particle energy, V is the potential energy, m is the particle's rest mass, c is the speed of light in vacuum, h is the Planck constant, and p is a quantity proportional to the nucleon density.

Knudsen Gauge

A device used to monitor very low gas pressures when the mean free path of the molecules is enormous in comparison to the size of the apparatus. After impacting electrically heated stationary plates at temperature T1, and then a cooler rotatable vane structure at temperature T2, gas molecules exert a torque on the vane. The pressure can be computed from the observed vane deflection 9 and equals $k\theta/((T_1/T_2)^{1/2} - 1)$, where k relies on the vane suspension's torsional constant.

Krypton

Symbol Kr. A gas utilized in the production of fluorescent lamps and laser. As a product of liquefaction of AIR, the gas is extracted from the atmosphere.

Kundt's rule

In the region of absorption bands, the refractive index of a medium does not fluctuate continuously with wavelength.

Kundt's Tube

An instrument used to measure the speed of sound. A small amount of dry powder, such as lycopodium, is dusted on the inside of tube T. Rod AB is clamped at its halfway, allowing the disc at B to just clear the tube's sides. The rod vibrates longitudinally when it is stroked with a cloth in the direction BA, producing a high-pitched note; disc B functions as a vibrating source of the same frequency, causing a sound wave to pass through the tube and be reflected back from end C. Adjusting the rod in the direction BC results in a position where the standing wave in the air in T violently agitates the lycopodium powder, causing it to settle into small defined heaps at the nodes; when the average distance between these is measured, it is equal to half a wavelength in air.

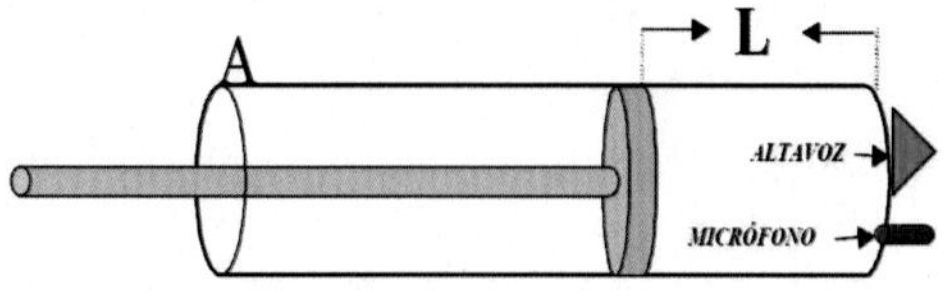

Figure 33. Kundt's tube.

Source: Image by Wikimedia commons.

L

Label

A radioactive atom in a molecule that is used to observe the molecule's behavior.

Ladder Filter

A chain of identical four-terminal symmetrical networks linked together to form a transmission line with continually repeated impedance portions.

Levorotatory

A substance that causes the plane of polarization of polarized light to rotate anticlockwise as observed by an observer facing the light source.

Lag

- The time interval or angle by which a certain phase in one periodically fluctuating quantity is delayed with regard to the same phase in another periodically varying quantity. There could be a lag between an alternating current and the electromotive force that generates it, or vice versa. Compare lead.
- The time elapsed between signal transmission and reception.

Lagging Load

A current-carrying load that lags behind the electromotive force that generates it. An inductor is a good example.

Lagrange's Equations

Second-order differential equations that represent the link between a particle system's Lagrangian function L, its generalized co-ordinates qi, its generalized forces Qi, and its time t:

$d/dr[dL/dq_i] - dL / dq_i = Q_i$ where $i = 1, 2, \ldots, n$ and n is the number of degrees of freedom of the system.

Lagrangian Function

Symbol L. A system's kinetic energy minus its potential energy.

Lalande cell

A primary cell with zinc and iron electrodes, a caustic-soda solution electrolyte, and conducts depolarization utilizing copper oxide.

Lambda Particle

A hyperon or an uncharged elementary particle with a mass 2183 times that of an electron. It has the ability to substitute a neutron in a nucleus, resulting in an exceedingly unstable hyper – nucleus.

Lambda Point

- The temperature 2.186 K at which helium types I and II are in equilibrium.
- The temperature at which a sharp maximum occurs in the specific heat capacity of a substance; for example, the temperature of the change from a ferromagnetic to a paramagnetic state.

Lambert

A luminance unit that is no longer in use, equivalent to the luminance of a surface radiating one light per square centimeter.

Lambert's Law

- The luminous intensity of a tiny element of a perfectly diffusing surface in any direction is proportional to the cosine of the angle formed by the normal and the direction.
- The temperature at which a sudden maximum in a substance's specific heat capacity occurs; for example, the temperature at which a ferromagnetic state changes to a paramagnetic state.

Lamb Shift

A tiny energy differential, caused by electron interaction with the radiation field, between the energy levels of the 2P and S states of hydrogen.

Laminar Flow

The continuous flow of a fluid in parallel layers with little or no combining between them. Compare with TURBULENT FLOW.

Lamination

A construction method utilized for the cores of transformers, transducers, relays, chokes, and other alternating current devices. The core is composed of thin strips of surface-oxidized or varnished iron or steel, which provides excellent resistance to eddy currents.

Lamy's Theorem

For a particle in equilibrium with three forces of magnitude A, B, and C, A/ sin a = B/sin = C/sin y, where a,~ and y are the angles between the lines of action of the forces B and C, A and C, and A and B, respectively.

Landau Damping

The dampening of a space charge oscillation via a stream of particles flowing at a slightly slower speed than the related wave's phase speed.

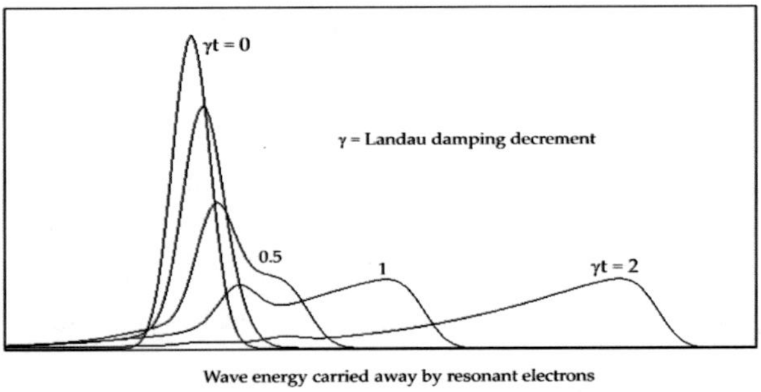

Figure 34. Landau Damping.

Source: Image by Wikimedia commons.

Lande Factor

Symbol g. A factor incorporated into the theory of electron energy variations in a magnetic field for attaining agreement with experimental observations like the anomalous Zeeman Effect. The values for g range from l for pure electron orbital momentum to 2 for pure spin momentum.

Lande interval rule

When the spin-orbit interaction is relatively weak, an atomic energy level separates into levels with intervals between them proportional to the bigger of the total angular momentum values.

Langmuir Effect

When atoms having low Ionization Potential interact with a hot metal with high work function, ionization takes place. The effect can be employed to generate powerful alkali metal ion beams.

Laplace Equation

- The equation $\nabla^2 V = d^2V/dx^2 + d^2V/dy^2 + d^2V/dz^2 = 0$, where (V) denotes a POTENTIAL and DEL squared (∇^2) denotes the Laplace operator or Laplacian; corresponding equations exist in spherical and cylindrical coordinates.
- The equation $c = (\gamma p/\rho)^{1/2}$, where c stands for the sound speed in a gas with density p, pressure p, and ratio of specific heats γ.

Larmor Precession

The precession of a charged particle's orbit when it is exposed to a magnetic field; the precession happens about the field's direction. The angular velocity of Larmor precession for an electron rotating around a nucleus is eH/(2mc), where c is the speed of light, H is the magnitude of the magnetic field intensity, and e and m are the electron charge and mass, respectively.

Laser

A powerful monochromatic coherent radiation source in the ultraviolet, visible, and infrared parts of the spectrum. The radiation is created via stimulated emission, with the resultant photon continuing the process until a narrow beam of monochromatic radiation is generated, if enough excited atoms are available. A condition known as population inversion, power must be supplied, for example, through optical pumping, in order to guarantee that there are enough excited atoms. Lasers that are gaseous, liquid, or solid have been developed. To accomplish population inversion, chromium ions are optically pumped, i.e., stimulated by an intense light flash, in the solid ruby laser, and stimulated emission is caused by spontaneous emission of wavelength 694.3 nanometer. The majority of the incident light is reflected back along the tube by full mirror, whereas the partial mirror reflects a portion of it.

Latent Heat

At fixed temperature, the total heat absorbed or produced during a phase shift.

Latitude

The angle formed by the plane of the equator and the normal to the Earth at the appropriate latitude.

Lattice

A two- and three-dimensional regular arrangement of points. A crystal system is formed when atoms, molecules, or ions are positioned at such points.

Lattice constants

The specified size and shape of a crystal structure's unit cell determined on the lengths of the cell edges and the sizes of their intersection angles.

Lattice Energy

The energy needed to split an ionic crystal into individual ions at an infinite distance apart at absolute zero temperature.

Lattice Vibrations

The regular vibrations of atoms, ions, or molecules about their mean locations in a crystal lattice. The amplitude of the vibrations grows with temperature until it reaches the melting point of the crystal and causes the lattice to collapse.

Laue Diagram

The diffraction pattern produced when an X-ray or other type of radiation or particle beam is delivered through a thin crystal and onto a photographic plate behind the crystal. The pattern can be used to determine the type of crystal and its structure.

Layer Lattice

A crystal lattice in which the atoms in the layers are firmly bonded but the bonding between the layers itself is poor. For instance, graphite has a layer lattice structure.

Lead Acid Battery

A form of accumulator made out of two lead plates that are immersed in dilute sulfuric acid. When both plates are discharged, they are coated with lead sulphate; when they are charged, the reaction is reversed. Each cell generates roughly 2 volts of electromotive force. A six-cell 12-volt battery is often used in automobiles.

Lead

- A conductor of electricity.
- The angle or time interval by which a specific phase of one periodically variable quantity is ahead of a similar phase of another such quantity. For example, there could be a lead between an alternating current and the electromotive force that generates it and vice versa. Compare LAG.

Lead Equivalent

The thickness of metallic lead that, under the identical conditions, would provide the same radiation protection as the substance under investigation.

Least Action Principle

The system action is minimal when a dynamic conservative system moves from one configuration to another.

Least Common Multiple

The smallest number which exactly divides each number in a given set of number. For instance, the lowest common multiple of 2, 3 and 5 is 30.

Least Distance of Distinct Vision.

The shortest distance between an object and the eye required for clear vision of the object. The typical figure for a young adult is about 25 cm.

Least Energy Principle

For a system to be in stable equilibrium, its total potential energy must be as low as possible.

Least Squares Method

The most probable value is determined by a number of observations on the premise that the sum of the DEVIATION squares of each value observed is a minimum value

Le Chatelier's Principle

When a system in balance has been changed, the system will change to counter the change. The idea has numerous applications, one of them being Lenz's law.

Leclanche Cell

A primary CELL having a cathode rod of zinc that is enclosed by a porous bag of manganese dioxide/graphite for avoiding polarization. A carbon rod as anode is employed, and the amount of ammonium chloride in the electrolyte is between 10% and 20%. The cell has an electromotive force of roughly 1.5 volts and is widely utilized in its dry form.

Lees Disc Method

A technique for determining the thermal conductivity of a poor conductor in the form of a thin disc having thickness d and face area A. The disc is between two metal plates and the top is heated via a steam chamber. The disc is heated. When the temperature of a thermometer in each slab becomes constant, the rate of heat flow through the disc matches the rate of heat loss from the bottom slab. This is determined by increasing the temperature of the bottom slab is increased above its previous value and plotting a cooling curve for it; the gradient d9/dt at the steady temperature of the bottom slab can then be calculated. The thermal conductivity is calculated as $(Msd\ d\theta/dt)/(A\ \delta\theta)$.

Lens

A transparent material with one or two curved surfaces, commonly glass, plastic or quartz. These surfaces are normally spherical (i.e., sections of spheres), though for special purposes they are utilized in other forms, such as cylindrical and parabolic. The numerous spherical converging lenses are the biconvex, plano-convex and concentric lenses, and these lens produce a real focus on rays of a parallel light beam, i.e., they converge the beam.

Lens Formula

The lens formula is the formula $1/f = 1/u + 1/v$, which (on the sign convention) real is positive and for rays that are close to and make small angles with the axis, pertains to thin spherical lenses; j stands for the focal length of the lens and u and v stand for the object and image distances from the lens, respectively.

Lepton

A grouping of elementary particles that do not engage in strong interaction. All leptons are fermions. The recognized leptons are the electron, muon, neutrinos, and tau leptons, as well as their antiparticles, which are the positron, positive muon, antineutrinos, and positive tau. Three quantum numbers known as lepton numbers, labelled 1_e, 1_u and 1_T, are assigned. The sum of the le values for all particles involved in the weak interaction is preserved, along with the sum of the 1_u and 1_T; values. Leptons lack a quark substructure.

Leslie's Cube

A huge cubic metal container featuring four vertical sides that are painted or finished differently on the outside. When filled with boiling water, it can be utilized to show the impact of surface nature on radiant heat emission by displaying each face in tum at a constant distance to a radiation detector.

Level

- The ratio of a quantity value to its reference value.
- The logarithm of a quantity value's ratio to a quantity's reference value.

Lever

A sturdy bar that can rotate around a pivot. The type is based on the relative positions of the load, effort, and pivot.

Levitation

The suspension of an object with no evident foundation. Electromagnetic levitation is created by creating eddy currents in an electric conductor with a time-varying magnetic field; it is the foundation of a cost-effective prototype transportation system.

Leyden Jar

A glass jar with tin foil coverings on the inside and outside. The inner coating is charged by connecting it to an appropriate voltage source, while the outside coating is grounded. The glass performs the function of a dielectric. The entire jar can be thought of as a parallel plate capacitor.

Lift Pump

A device where an external handle moves a piston holding a valve V up and down a cylinder. Another valve V' at the cylinder's base connects to a pipe whose other end is in the liquid to be elevated. When the piston is raised, atmospheric pressure pulls liquid into the cylinder beyond V'; when the piston is lowered, water is driven through V and delivered from the top of the cylinder on the next upstroke. The mechanism is commonly used to raise water from wells that are shallower than the maximum depth of water (about 10 meters) that air pressure can support.

Light

Electromagnetic radiation that causes a visual sensation upon reaching the human retina. Light has a wave length range of 400 nanometers to 740 nanometers. Many theories have been proposed to explain the nature of light, including corpuscular theory, quantum optics, and quantum electrodynamics.

Light Emitting Diode

A rectifying semiconductor device that is different from a standard semiconductor diode as its electron hole recombination mechanism results in light emission. The given figure depicts the electrical connections and physical assembly of a light emitting diode. The device is forward pushed, and the current is limited by a resistance; light is emitted through the narrow p-type region, which has the least amount of obstruction. The color of the light depends on the semiconductor material used: a red or yellow light is emitted by gallium arsenide phosphide, whereas a green light is emitted by gallium phosphide. No time lag in the production of light makes rapid switching feasible. Light-emitting diodes are frequently employed in self-luminous digital equipment to display characters and numbers.

Lightness

A color's property defined by the quantity of light it reflects. Shades are colors that have the same hue but vary in lightness.

Lightning

Forked lightning is a high-energy luminous electric discharge that occurs between a charged cloud and a point on or linked to the Earth's surface. Sheet lightning takes place when a discharge occurs between two charged clouds or between oppositely charged layers of the same cloud. A potential difference of around 10^8 volt is necessary to start a flash. In general, there is a downward leader stroke, which is a partial discharge, followed by an upward return stroke, which is significantly brighter.

Lightning Conductor

A sharply pointed metal rod connecting to the Earth's surface and attached to the top of a building. By induction, a sharp point in the vicinity of a charged cloud becomes significantly charged, ionizing the air above it. As a result, a discharge between the cloud and the conductor occurs more slowly and less strongly than a discharge between the cloud and the Earth.

Light Pen

A device that is linked to a computer's visual display unit. It can detect information on the screen and can be used to convey information by 'drawing' lines on the screen.

Light-Sensitivity Materials

Substances that absorb visible and ultraviolet photons and produce excited electronic states or electron emission as a result.

Light Year

A unit of length that gives the distance travelled in one year by electromagnetic radiation in vacuum, i.e., 9.4607×10^{15} meters, It is a unit of measurement for astronomical distances.

Figure 35. Light year.

Source: Image by Max Pixel.

Limit

The value approached by a mathematical function when its independent variable approaches a predefined value.

Limiting Angle of Prism

The maximum angle at the refracting edge of a prism of a particular transparent material for which an emergent ray can be attained. Its value is approximately double the prism material's critical angle.

Linde Process

A method of liquefying air that involves expanding it through a nozzle. The cooled air is utilized for cooling the entering compressed air, eventually reaching liquefaction temperature.

Linear Accelerator

A particle accelerator that accelerates electrons or protons down a straight evacuated chamber. Earlier, particles were accelerated in a drift tube, which was a series of cylindrical electrodes separated by gaps: a radio frequency electric field was applied to the cylinders having such lengths that the particles were accelerated at each gap, with the velocity remaining unchanged until the next gap. Modern high-energy machines are mainly travelling wave accelerators rather than drift tubes: particle acceleration is created by the electric component of a travelling wave established in a waveguide; energy gain rates for electrons are typically 7 mega-electron-volts per meter and 1.5 mega-electron-volts per meter for protons.

Linear Attenuation Coefficient

Symbol μ · A coefficient given by $(d\phi/dl)/\ \phi$, where $d\phi/dl$ is the rate of change of energy flux ϕ, across distance l. Due to absorption and dispersion by the medium, the flux changes as it travels through it.

Linear Energy Transfer

The energy transferred to atoms and molecules along a charged particle's route per unit of travel length. Linear energy transfer is proportionate to the square of the particle charge and multiplies as particle velocity decreases. The notion is most commonly applied in radiation protection efforts.

Linear Momentum

Symbol p. A vector quantity determined by the product of a particle's mass and velocity. The linear momentum of a body or particle system refers to the vector sum of its individual components' linear momenta.

Linear Motor

An induction motor where the stator and rotor are linear and parallel instead of cylindrical and coaxial, like in a standard rotary machine. The motor's objective is to generate force or motion in a straight line.

Line Frequency

The number of lines scanned by an electron beam in a television receiver's cathode ray tube every second.

Line of Force

An imaginary line with a direction at any point along its length that corresponds to the direction of the force field at that point. Electric lines of force, magnetic lines of force, and gravitational lines of force are some examples.

Line Printer

A device that prints a computer's output one line at a time rather than individual characters. Line printers can print at speeds of up to 50 lines per second.

Line Spectrum

A spectrum made up of discrete lines. These spectra are produced by photon emission or absorption caused by electron transitions between different energy levels in atoms.

Figure 36. Line spectrum.

Source: Image by Wikimedia commons.

Liquefaction of Gases

The conversion of a substance from a gaseous to a liquid state. When the gas temperature is under the critical temperature, pressure alone can be used to liquefy it; else, the Linde process, adiabatic expansion, or adiabatic desorption can be used.

Liquid

A phase of matter that exists between a gas and a solid and is distinguished by its ease of flow and incompressibility. A liquid has negligible shear stress resistance and adapts to the shape of its container; contrary to a gas, it does not alter volume to occupy the container. Liquids have stronger intermolecular forces than gases but weaker forces than solids. Diffraction studies reveal that liquids have a short-range structural regularity that spans several molecule diameters; these ordered bundles move relative to one another. Liquid theories are not as researched as solid and gas theories.

Liquid Crystal

A substance with the flow qualities of a liquid but a more organized molecular arrangement than a liquid. Liquid crystals are formed by lengthy molecules, the arrangement of which determines the type of liquid crystal. An electric field affects the polarizing properties of liquid crystals, which have the property of plane polarizing light.

Liquid Crystal Display

A digital display device that consumes extremely little power and is found in watches and calculators. A thin sheet of liquid crystal is sandwiched between transparent electrodes, each of which is glass thinly covered with metal. The design on the film generates the desired display. The polarization planes of the polarizers (not visible) on either side of the liquid crystal display are intersected.

Liquid Drop Model

A nucleus model wherein the nucleons are shown as liquid molecules and the entire nucleus by a liquid drop. The shape of the nucleus is thought to be maintained by the forces of interaction between surface nucleons, just as the shape of a drop is maintained by surface tension due to the interaction of molecules at the surface. The model, which is utilized in nuclear fission theory, is best suitable to heavy nuclei.

Lloyd's Mirror

A plane mirror used for generating interference fringes: monochromatic light from a slit is grazed by the mirror and then impedes with direct light coming from the slit.

Loaded Concrete

Concrete containing high-atomic-number materials, which include barium, iron, or lead. Its principal application is as a radiation shield in nuclear reactors.

Locus

A curve traced by a moving point in order to satisfy a specific condition. A circle, for example, is the locus of a point that moves in a way that its distance from a fixed point, the center, remains constant.

Logarithm

The power to which a number (the base) must be raised in order to equal a specific integer. Any number x, for example, can be expressed as y^n, where n is the logarithm to the base y of x, i.e., $n = \log_y x$. When y = 10, the logarithms are referred to as common logarithms. When y = e = 2.71828..., the logarithms are known as natural logarithms, hyperbolic logarithms, or Naperian logarithms, and are denoted by the symbols $\log_e x$ or lnx.

Logic Circuit

A circuit that is meant to execute a specific logical function, such as 'and,' 'either... or,' or 'neither... nor.' The circuits are typically binary logic circuits, which operate between two distinct voltage levels; these circuits are widely used in computers and are typically created from an integrated circuit assembly. A logic network refers to a group of logic circuits.

Longitude

The angular distance between the Earth's meridian at the location in question and the standard meridian, which is a large circle that runs through the poles and Greenwich and whose longitude is ascribed the value 0°.

Longitudinal Wave

A wave in which the transmitting medium's vibrations run parallel to the wave's travel direction. A sound wave is an example.

Loudness

Magnitude of the sensation caused by a sound entering the ear. The relationship between loudness and intensity varies based on the frequency and duration of the sound; however, loudness is roughly proportional to the logarithm of intensity.

Loudspeaker

An instrument that converts varying electric currents into sounds. In its most popular configuration, a tiny coil of wire is connected to a cardboard cone, with the coil located in the field of a powerful permanent magnet. As a result, when alternating current flows through the coil, it vibrates. These vibrations are conveyed to the cone, resulting in sound. The gadget should respond to frequencies ranging from 150 hertz to 8000 hertz for optimal speech reproduction. Response in the 20 hertz to 20 000 hertz range is essential for optimal music reproduction.

Lubrication

The process of inserting a substance between two solid surfaces that are moving in relation to each other in order to reduce friction, wear, overheating, and corrosion. Different types of oils and greases are extensively utilized, however at high temperatures, graphite is preferred. Air bearings are becoming more popular, although they require continuous air pumping to the bearings.

Lumen

Symbol lm. The SI unit of luminous flux, equivalent to the luminous flux generated by a uniform point source with an intensity of one candela in a cone with a solid angle of one unit.

Luminance

Symbol L_v. The product of a surface's luminous intensity per unit area when viewed from a specific direction and the secant of the angle 9 between the surface and that direction. Hence, $L_V = \sec\theta\, dl/dA$, where l is the luminous intensity and A is the surface area. The unit of measurement is candela per square meter.

Luminescence

The release of electromagnetic radiation from a substance due to any nonthermal action. Electron transitions in excited atoms and ions cause luminescence. Electromagnetic radiation, electronic bombardment, biological responses, friction, radioactive decay, and chemical reactions can all cause excitation; the related types of luminescence are photoluminescence. Electroluminescence, bioluminescence, triboluminescence, radioluminescence and chemiluminescence, respectively. Fluorescence or phosphorescence radiation may accompany the return of electrons to the initial state.

Luminosity Curves

Plots of the variance of the eye's reaction to an equal energy spectrum as a function of wavelength. The curve at low levels of illumination or the

Scotopic Vision curve is not the same as that at high levels of illumination or the Photopic Vision curve. The maxima of the curves are separated by approximately 0.5 x 10^{-7} meter. Even though photopic vision requires far higher intensities than scotopic vision, the curves depicted have been adjusted to the same maximum height.

Luminous Paint

Paint containing a phosphor combined with a trace of radioactive substances. Light is radiated because of the phosphor being bombarded by radioactive material releases.

Lummer-Brodhun Photometer

A photometer. Because of the optical system employed, the illumination from the test source appears as a spot surrounded by a zone illuminated by the standard. The sources' placements are changed until the area of vision looks uniformly lit.

Lunar Time

Time calculated in relation to the Moon. The lunar month is the period between consecutive new moons; a lunar year is made up of 12 lunar months, which is equivalent to 354.3671 mean solar days.

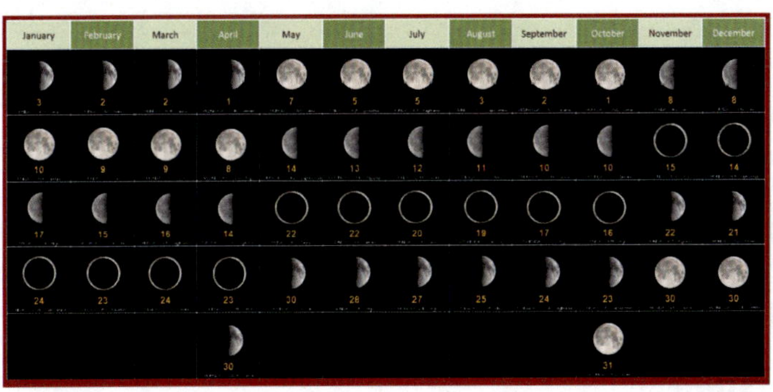

Figure 37. Lunar Time.

Source: Image by Wikimedia Commons.

M

Mach Angle

The cone's semi-angle, or the envelope, i.e., the wave front, of spherical pressure waves produced by a body traveling at supersonic speed through a medium. It is expressed as sin – 1(1/M), where M denotes the Mach number.

Machine

A system that allows a mass, or load, to be moved with a significantly smaller force, or effort, than the weight of the load. The mechanical advantage of the machine is the ratio of load to effort, and the velocity ratio is the ratio of distance moved by effort to distance moved by load. The machine's efficiency is defined as the ratio of work done on the load to work done by effort. It is frequently given as a percentage and equals the ratio of mechanical advantage to velocity ratio: it cannot be more than100%, which is the value for a frictionless or a flawless machine.

Mach Number

Symbol M. The ratio of a body's relative velocity through a fluid to the local velocity of sound in the fluid. If M is greater than one, the relative velocity is supersonic whereas if M is greater than five, the relative velocity is hypersonic.

Macroscopic

(1) Large enough to be seen without a microscope.
(2) Having traits related to the statistical behavior of a huge number of atoms or molecules.

Magic Numbers

The numbers 2, 8, 20, 28, 50, 82 and 126. They get their names from the fact that atomic nuclei with this number of protons or neutrons are more stable than other nuclei. The numbers correlate to the nucleus's filled shells. Tin and calcium, for example, are magic number elements and have a comparatively significant number of isotopes.

Magnet

A device that generates a magnetic field. A magnet can be temporary or permanent – an electromagnet's magnetic field exists only when current flows through it, while a magnetized piece of ferromagnetic material has a permanent magnetic field which makes it a permanent magnet.

Magnetic Balance

A device for determining the force between magnetic poles directly. A long bar magnet is balanced horizontally on a knife edge in a basic form. Another pole is then positioned above or below one of the magnet's poles, deflecting the latter downwards. The force between the poles can be calculated using the location of a rider that can be moved along the bar magnet to reestablish horizontal balance.

Magnetic Bottle

A magnetic field configuration used to contain plasma in controlled fusion, mostly in a linear container.

Magnetic Constant

Symbol Jo. The ultimate permeability of free space. It is equal to 4n x 10^{-7} henry per meter.

Magnctic Cooling

A technique for obtaining extremely low temperatures. Isothermal magnetization of a paramagnetic salt at low temperature proceeded by adiabatic demagnetization can result in a temperature of 10^{-3} K. If isothermal magnetic alignment of a substance's nuclear spins with nuclear magnetic moments is followed by adiabatic demagnetization, at a temperature of roughly 10^{-2} K, a temperature as low as 10^{-6} K can be obtained.

Magnetic Crack Detection

A method for detecting discontinuities on or near the surface of a ferromagnetic material. The substance is coated with a dispersion of small iron particles in oil, and the magnetization distribution is studied: irregularities reveal the locations of discontinuities because of the cluster of iron particles formed there.

Magnetic Cycle

A plot of magnetic induction magnitude against magnetic field strength.

Magnetic Dipole

A north and south pole having equal force separated by a distance. In contrast to a positive or negative electric CHARGE, which can exist independently, each magnetic pole is assumed to exist as one member of a dipole.

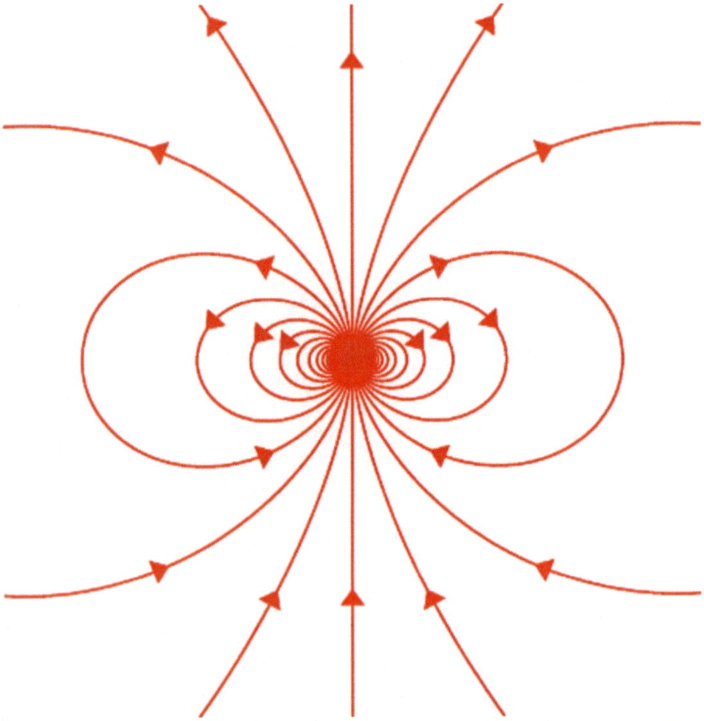

Figure 38. Magnetic dipole.

Source: Image by Wikimedia Commons.

Magnetic Dipole Moment

The product of the pole strength of one of a magnetic dipole's two poles and the pole spacing. A tiny wire loop of area A with a current I operate as a magnetic dipole and has a magnetic moment lA, commonly referred to as the electromagnetic moment.

Magnetic Domains

Small zones of intense magnetism appearing within purportedly unmagnetized ferromagnetic material.

Magnetic Equator

A line drawn on the Earth's surface that connects all zero DIP spots. It can be found anywhere near the equator.

Magnetic Field Strength

Symbol H. A vector quantity given as the magnetic induction divided by the medium's absolute permeability. The magnitude is expressed in amperes per meter.

Magnetic Flux

The product of the area under study and the component of the average magnetic induction over it that is normal to the area. Therefore, it is given as the surface integral of magnetic induction normal to the surface.

Magnetic Induction

Symbol B. A vector whose magnitude is defined by F/(qv sin 9), where F is the magnitude of the force endured by a moving charge q travelling with velocity magnitude v in a direction making an angle 9 with the field direction, and whose direction at a given position is that of the magnetic field at that point. The magnitude of magnetic induction is expressed in the unit tesla.

Magnetic Meridian

A large circle travelling through the point on the surface of the Earth under examination as well as through the Earth's magnetic poles.

Magnetic Mirror

A magnetic field strong enough and properly oriented to redirect the direction of charged particles. In a magnetic bottle, magnetic mirrors are employed.

Magnetic Pole

A region located near the poles of a permanent magnet. The magnetic field's force lines converge on one pole and diverge from the other. A freely suspended magnet comes to a stop with its north-seeking pole facing north and its south-seeking pole pointing south in the direction of the Earth's magnetic field. Magnetic poles that are unlike attract one other, while those that are like repel each other. The force, much like force between electric charges, abides the inverse square law of variation with distance.

Magnetic Recording

The recording of sound on magnetic tape and subsequent reproduction of the sound from the tape. The tape runs at a consistent pace through a very thin air gap in a ring of soft iron that is magnetized by the recording amplifier's current going through the coil surrounding the ring.

Magnetic Screening

The shielding of a region from magnetic effects by enclosing it in a material having a strong relative permeability.

Magnetic Shell

A thin sheet of ferromagnetic material magnetized all the way through. It is equivalent to an endless number of tiny magnets.

Magnetic Storm

A disruption in the magnetic field of the Earth caused by an electrical disturbance induced by sunspot activity.

Magnetic Susceptibility

The magnetic dipole moment per unit volume divided by the magnetic field strength.

Magnetic Tape

A plastic tape that has been covered with iron oxide and can be magnetized to record sound. Another use is the storage of binary information as magnetic dots used in computers.

Magnetomotive Force

The magnetic equivalent of electromotive force. It is the line integral of H cos 9ds around a closed route, where H represents the amount of the magnetic field intensity and e represents the angle it makes with the path element ds.

Magnetopause

The magnetic field's outer border on Earth. It is about ten times the radius of the Earth from its center.

Magnetoresistance

A change in a substance's electrical resistivity caused by the presence of an external magnetic field. It can be found in all metals, but it is most noticeable in ferromagnetic metals and their alloys.

Magnetron

An electron tube used to generate microwaves. A hot cathode in the tube is surrounded by a coaxial cylindrical anode. When a significantly strong magnetic field is applied parallel to the cylinder's axis, the electrons generated by the cathode are redirected towards it and revolve around it. This causes the induction of radio frequency fields in the anode's resonant cavities; these fields operate on the electrons as well, causing oscillations.

Magnifying Power

The ratio of the angle subtended at the eye by an optical system's final image to the angle subtended by the object at the naked eye.

Magnitude

A celestial object's brightness, with the brightest objects having the least magnitude. A brightness ratio of 100 to 1 corresponds to a magnitude difference of 5, therefore a star with magnitude m is 1000.2, or 2.512, times brighter than a star with magnitude m + 1.

Magnon

A quantum of magnetic energy that is equivalent to a photon. It is useful when determining the excitation or de-excitation of magnetic spin waves by neutrons.

Magnox

Any custom magnesium alloy utilized in some nuclear reactor types to encapsulate fuel components. Magnox A, for example, comprises 0.8% aluminum and 0.01% beryllium.

Many-Body Problem

The problem of solving the equations defining the interaction of a number of free particles with given initial locations and velocities and subjected to given forces. In general, no rigorous solution can be obtained for more than two bodies. Planetary motion, many-electron systems, and nucleon interactions are all examples of many-body problems.

Mars

In terms of increasing distance from the Sun, the fourth planet in the solar system. It has a tenth of the mass of Earth and a half-diameter. It has an axial rotation period of around 24.5 hours and an orbital period of 687 days. The planet has two satellites and is reddish in color. Explorations to mars by space probes have been done.

Maser

An amplifier or oscillator that functions on the same principles as the LASER and exhibits the same features, but in the microwave area of the spectrum. The term is an abbreviation for microwave amplification by stimulated emission of radiation.

Mass

Symbol m. The amount of matter in a body. It quantifies a body's propensity to resist a change in motion. As per Newton's laws of motion, force applied is proportionate to acceleration produced, with mass serving as the proportionality constant. The mutual attraction of two bodies due to gravitational interaction is also determined by mass. According to relativity theory, mass fluctuates with velocity and is interconvertible with energy.

Mass Defect

The difference in mass between the sum of the masses of individual nucleons in a nucleus and the mass of the nucleus itself. The energy binding the nucleons together is equivalent to the difference in mass.

Mass-Energy Relation

A relationship defined by the equation $E=mc^2$, where E is the amount of energy generated by a mass change of m and c is the speed of light in free space. Hence, matter can be generated or destroyed, resulting in a proportional loss or gain in energy. Although the conservation law for mass and energy evaluated separately is not technically valid, the total mass-energy is preserved.

Mass Spectrometer

A device for determining the relative abundance and mass of isotopes in a sample. In most cases, the sample is used for coating a filament. When the filament inside a gas discharge tube is heated, atoms from the specimen are ejected into the tube, where they are bombarded with electrons and therefore become ionized. These ions pass via a narrow slit and into a velocity selector. The selector is strongly evacuated and exposed to right-angle electric and magnetic fields, allowing only ions of constant velocity to exit through the end slit.

Figure 39. Mass spectrometer.

Source: Image by Flickr.

Maximum and Minimum Thermometer

A thermometer from which the maximum and lowest temperatures reached from the time the thermometer was set may be read. Each branch of a U tube comprises a steel indicator I. Rise in temperature causes the alcohol in A to expands, leading to the mercury in the U tube to rotate, pushing the indicator on the maximum side up the tube.

Fall in temperature causes the alcohol condense, but the maximum indication remains in place due to the action of the little spring attached to it; when the mercury thread condenses, the indicator on the minimum side rises as well. A permanent magnet can be used to reset the indicators.

Maximum Density of Water

At a temperature of 4°C, water has a maximum density of 1000 kilograms per cubic meter. This is because ice crystals have a fairly open three-dimensional tetrahedral structure, a structure which is not entirely replaced by the closer-packed water molecular structure till the temperature exceeds 4°C.

Memory

The component of a computer that saves data in *word* or *byte* units for instant use by the central processing unit. Solid-state memory is now the norm, having replaced magnetic core storage. Solid-state memory is smaller and faster than core store memory. It could be RAM (random access memory) or ROM (read-only memory), both of which are made in the form of an integrated circuit. ROM's contents are locked, whereas information in the RAM can be written and read from. When the supply voltage is disconnected, the information in RAM is erased.

Mercury

The planet closest to the Sun. It has a mass of roughly 0.05 and a diameter 0.38 times that of Earth. It has an axial rotation period of 58.6 days and an orbital period of 88 days. Mercury is a challenging star to view and can only be seen in the morning or sunset.

Mercury Cell

A cell with a cathode made of compressed mercury II oxide and graphite. In a concentrated potassium hydroxide solution, either zinc or cadmium interacts with the mercury II oxide to produce mercury. The cell energy density is high because the electrolyte is not consumed. This allows it to be made tiny enough for use in calculators and other devices. Moreover, it is ideal for pacemaker applications as it has stable voltage, strong current supply without performance degradation, and a long shelf life.

Meson

Any elementary particle with integral spin, strong interaction, and a mass halfway between that of a proton and that of an electron. A meson can be neutral, positively or negatively charged, and have the same charge magnitude as an electron; positive and negative mesons are antiparticles of each other. Mesons are in charge of the forces that exist between nucleons in the nucleus. All are unstable and degrade in various ways. It has been proposed that a meson is made up of a firmly bound quark and an antiquark.

Metacenter

The point at which a vertical line passing through the center of buoyancy of a tilted floating body intersects the line connecting the center of mass with the position of the body's center of buoyancy when upright. As long as the meta-center is higher than the body's center of mass, the body will continue to revert to a vertical position.

Metal-Glass Seal

A sort of seal that is needed in various pieces of equipment; usually, vacuum tightness is required. It is vital to make the seal when the glass is soft, i.e., at a high temperature, and hence glass and metal with nearly the same expansion coefficient are preferably required to prevent cracking on cooling. For small tasks, soft glass and copper-coated nickel-iron alloy is an excellent mix. When making larger joints, such as between glass and metal tubes of several centimeters diameter in a vacuum plant, a tapered copper tube is used: the copper coming in contact with the glass is then slim enough to endure the distortion caused by the difference in expansion coefficients without cracking the glass.

Metallic Bond

A metallic crystal BOND in which a regular lattice of positive ions is kept together through a cloud of free electrons circulating through the lattice.

Metallizing

The procedure of coating an insulating material with a metal film to make it electrically conductible. The approach is vital in solid-state electronics.

Metallurgy

The science of metals and their alloys, covering their design, characteristics and industrial smelting, refining and working procedures.

Metal Rectifier

A metal plate in contact with its oxide or another suitable compound that provides significantly more resistance to current flow in one direction than in the opposite direction; examples include the copper oxide rectifier and the selenium rectifier. Such equipment cannot be used for potentials exceeding a few volts; therefore, a series of corrective devices are used for high voltages. The transmitted current is based on the contact area.

Meteor

A particle of matter that reaches the Earth's atmosphere from space. The composition is varying, although iron and nickel alloys are found abundantly. The surface temperature of a meteor reaches around 3000 K causing the meteor to seem brilliant due to friction from air particles. A meteor usually leaves a trail of ionized gas behind.

Meteorite

A massive METEOR, a portion of which hits the Earth's surface, the remaining being burned off in its journey through the atmosphere. The mass reaching the Earth's surface can range from a few grammes to 65 000 kilograms, as discovered in Grootfontein, South America.

Method of Mixtures

A calorimetric procedure that involves placing a given mass of liquid at a given temperature in a lagged calorimeter. A specified mass of another substance at a different known temperature is added to the liquid, which is then agitated and its temperature is recorded. If the values of all variables but one is known, the value of that variable can be estimated using the principle that heat released by hot bodies is equal to the heat received by cold bodies.

A few common determinations are: (a) specific heat capacity of a solid – the heated solid is transported to the liquid in the calorimeter; (b) specific heat capacity of the calorimeter – a heated solid with known specific heat capacity is transferred to the liquid in the calorimeter; (c) specific heat capacity of a liquid- the liquid is used in the calorimeter and a solid with known specific heat capacity is added; (d) Specific latent heat of ice – dry ice at 0°C is transmitted to water in the calorimeter; (e) specific latent heat of steam- dry steam at atmospheric pressure is transferred to water in the calorimeter

Meter

Symbol m. The SI unit of length, as given in October 1983 as the length of light's passage in vacuum over a time span of 1/299,792,458 of a second. It was formerly equal to 1,650,763.73 wavelengths in free space of the radiation correlating to the transition of the krypton–86 atom's $2p_{10}$ and $5d_5$ levels.

Mica

A material made up of complex silicates that easily breaks into very thin plates. Because of its low thermal conductivity and strong dielectric strength, it is commonly used as an electrical insulator.

Microbalance

A beam balance that can measure masses as minuscule as w-s gramme. One end of the beam is fixed with a bulb. The force of buoyancy on the bulb is changed by adjusting the air pressure in the balancing case. At constant temperature, this force is proportional to the pressure, which can be calculated with a manometer at balance.

Micromanometer

A gadget that measures very small pressure variations. The diaphragm gauge, for example, calculates the diaphragm displacement caused by the pressure difference optically.

Micrometer Eyepiece

A Ramsden eyepiece with cross wires that may be shifted by a micrometer screw, allowing minuscule distances to be properly measured.

Micrometer Screw

An instrument for measuring distances accurately. It uses a calibrated drum that, when rotated, advances or retracts a screw of known pitch; the distance moved may be properly determined by examining the drum readings.

Microphone

A device that converts sound energy into electric energy. It is found in telephones, sound recorders, and broadcasting equipment, among other things. Various energy transformation methods have been devised, culminating in variations in resistance, capacitance, electromagnetism, temperature, piezoelectric effect, magnetostriction, or electrostriction. The objective is to guarantee that the electric and acoustic vibrations coincide as nearly as possible to provide realistic sound reproduction in the end.

Microscope

An optical device that produces a magnified image of a nearby object. A simple microscope is a converging lens system that has been corrected for chromatic and spherical aberration. The object is typically placed at a distance smaller than the focal length of the lens, resulting in a virtual erect magnified picture at the shortest distance of distinct vision.

The magnification is equivalent to the magnifying power. A compound microscope is a device that consists primarily of two convex lens systems with short focal lengths. The object is located at a distance from the objective that is slightly greater than its focal length, giving a true enlarged inverted image. This image is at a distance from the eyepiece that is slightly less than its focal length, resulting in an erect enlarged virtual image of the first image at the least distance of distinct vision.

Microwave Background

An isotropic dispersion of microwave radiation across the universe. It coincides with black body radiation having a temperature of 2. 7 K and is very certainly a relic of the big bang, which is thought to be the universe's origin.

Microwave Spectrum

A spectrum of emission or absorption in the microwave range. Data about molecular size and moments of inertia can be derived from such spectra.

Middle Ear

A tiny air-filled hollow in the ear that contains three articulated bones known as the ossicles, which are also known as the hammer, anvil, and stirrup. Their role is to transport pressure changes from the eardrum to the middle ear; the dimensions ensure that the pressure at the eardrum is amplified between 40 and 90 times throughout the transmission process.

Mie Scattering

Light scattering by spherical particles with sizes corresponding to the wavelength of the light It is derived from Rayleigh Scattering.

Mirage

An effect generated by the progressive reduction in air density as one approaches a hot ground surface. Light from the sky is finally completely internally reflected at a sufficiently low-density air layer, creating the illusion of a pool of water. A tree in a hot desert is frequently viewed as its reflection in a pool of water.

Mirror

A polished barrier separating two optical media at which light is reflected. In general, silvering or aluminizing the boundary surface enhances the brightness of the reflection.

Mirror Galvanometer

A galvanometer, whose coil's deflection is ascertained by the deflection of a spot of light reflected from a mirror coupled with the coil.

Mirror Image

An object's image as it would look in a plane mirror. If one object is the mirror image of the other, then they are identical except that they will not superimpose in most cases. A right hand, for example, is the mirror image of a left hand.

Mobius Strip.

A surface created when a band is twisted through 180° and then the ends are connected. As it is a single surface having a single bounding curve, it is mathematically interesting.

Moderator

A material, generally a light element like graphite, used in a nuclear reactor to slow down free neutrons to make them more likely to generate FISSION in uranium–235 rather than being absorbed by uranium–238.

Modulation

The process of imposing one wave system, namely the signal, on another wave system, namely the carrier, in order for the signal's information to be transferred by the carrier wave.

Molar Polarization

The shift of electrical centers as a result of a molecule being subjected to an electric field, leading to the formation of a DIPOLE in the molecule. The polarizability of a molecule is defined as the ratio m/E, where m is the dipole moment and E is the magnitude of the electric field strength.

Mole

Symbol mol. The fundamental SI unit of amount of substance. It is the quantity of any substance having the Avogadro Constant of entities such as atoms, molecules, ions, electrons, and so on.

Molecular Beam

A low-pressure collimated beam of atoms or molecules. There are rarely intermolecular collisions since all of the particles are travelling in the same direction. A particle stream is fed through many apertures to create such a beam, with particles that do not pass through being evacuated by a vacuum pump. In spectroscopy, molecular beams are used.

Molecule

The smallest grouping of atoms that can be thought of as a chemical unit of structure. For ionic crystals and many complex organic molecules, the idea is of little relevance; for the latter, the crystallographic unit is more relevant.

Moment

The impact of a force rotating about an axis. The magnitude of the moment is calculated as the product of the magnitude of the force and the perpendicular distance from the axis to the force's line of action. If the algebraic total of the moments acting on an item about any axis is zero, the object is in rotational equilibrium. This can also be stated as the sum of the clockwise moments about any axis equals the sum of the anticlockwise moments about the same axis at equilibrium.

Moment of Couple

Product of the magnitudes of any of the forces that make up a couple and the perpendicular distance between their lines of action.

Moment of Inertia

Symbol I The sum of the product of a particle's mass and the square of its perpendicular distance from the axis under concern, for all particles in a body; for a homogeneous body, the total is calculated through integration.

Monitoring Pill

An extremely miniature integrated circuit or circuits embedded in a pill. When a patient swallows the tablet, it continuously communicates information about the conditions in the patient's tracts. An external receiver monitors the transmissions, which is highly beneficial in diagnosing situations that might otherwise create diagnostic problems.

Monochord

A thin metal wire strung horizontally over two bridges. The stretching force is exerted by attaching weights to one end of the wire and running the other end via a pulley attached to the hollow box on which the sonometer is mounted. The string can be made to vibrate by striking, bowing, or plucking it. The length can be adjusted by repositioning the bridges. The hollow box serves as a sounding board, enhancing the amount of sound produced by vibrations. The setup can be used to test the relationship between a wire's length, diameter, tension, and vibration frequency.

Moon

The sole natural satellite of Earth, revolving at a mean distance of 384 400 kilometers every day in around 29.5 days. The diameter of the Moon is 3476 kilometers, and its relative density is 3.34. It has virtually no atmosphere and no surface water. The minimum night temperature is around 80K, and the highest day temperature is around 400K. The Moon's gravitational force has a significant impact on Earth's tides. Humans first set foot on the Moon in 1969.

Moving Coil Instrument

An electric measurement device. An electromagnet's flat rectangular coil of wire is positioned vertically between its curving pole pieces and it conducts direct current. As a result, the vertical sides of the coil are subjected to opposing forces, forcing the coil to rotate until it is restrained by the torsion in the suspension. The angle rotated through is proportional to the size of the current and is quantified by the deflection of a light beam reflected at the small mirror attached to the suspension; the device is thus referred to as a moving coil galvanometer.

N

Natural Dosage of Radiation

The radiation dosage because of natural radio, cosmic rays, active materials in rocks and soil, and small amounts of radioisotopes, chiefly potassium–40, arising naturally in the body. The total amount of natural dosage per annum from all sources is 1.25 x 10^{-3} sievert: 0.5 x 10^{-3} sievert from cosmic rays and the identical from the Earth and 0.25 x 10^{-3} sievert from potassium–40.

Near Infrared and Ultraviolet

The portions of the spectrum of electromagnetic radiation nearby to the visible region. Near infrared emission and absorption are generally linked with transitions between vibrational energy levels of a molecule.

Near Point

It basically refers to the closest point to the eye at which clear vision is possible.

Nebula

A cloud of interstellar dust and gas. Majority of them are clearly observable from the naked eye. Some are largely visible by their emitted radiation and others by the radiation they imitate; some absorb radiation.

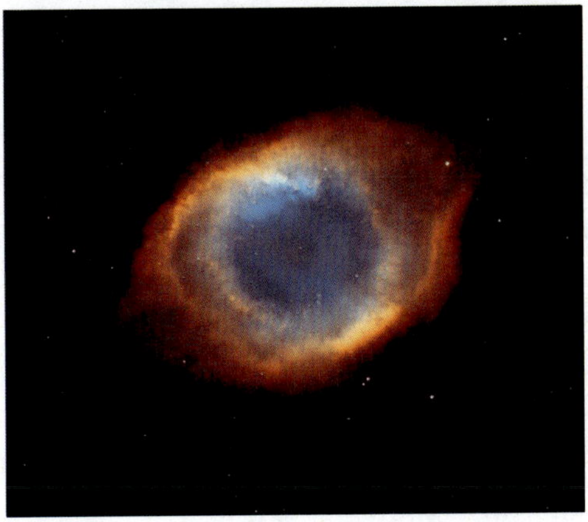

Figure 40. Helix Nebula.

Source: Image by Flickr.

Nebular Hypothesis

Any of a group of theories contingent upon the assumption that the solar system initiated from the condensation of a nebula.

Negative Resistance

The explanation of the negative slope of some voltage against current plots. Some of the available devices that shows this phenomenon are magnetrons, silicon – controlled tunnel diodes and rectifiers.

Negative Specific Heat Capacity

The specific heat capacity of a substance from which it is essential to extract heat in order to surge its temperature. A saturated vapor is an instance.

Neon

Symbol Ne. A gas that can be attained from the atmosphere as a byproduct in the liquefaction of air. It can be used in fluorescent lamps and neon tubes.

Neon Tube

A gas discharge tube comprising neon at low pressure; it glows red. The voltage across a tube remains persistent over a range of currents and so, for currents up to 100 milliampere, a tube can be used as a voltage stabilizer. Electrodeless tubes eagerly glow in the existence of high voltage high-frequency currents and so may be used for the detection of such currents.

Neptune

The outermost giant planet. distance 4.5×10^6 kilometer from the Sun. Its diameter is 48,400 kilometer, its mass 17.2 times that of Earth, its axial and orbital rotation times 164.79 year and about 16 hour correspondingly. The temperature is around –205°C and the atmosphere comprise largely hydrogen and methane.

Nernst Calorimeter

An apparatus, denoted by symbol; and it is used to measure the particular heat capacity of a metal specimen S, mass m. C is a coil that can be electrically heated; F is a layer of foil to curtail radiation heat losses. Conduction and convection heat losses are eradicated by emptying the enclosure. The amount of heat supplied is IVt where I is the current flowing through C, t the time for which it streams and V the voltage across C.

Nernst Effect

The formation of a potential distinction mutually perpendicular to a semiconductor or conductor along which there is a temperature gradient, and to an applied magnetic field. The effect is analogous to the hall effect.

Net Radiometer

An instrument for measuring the variance in intensity between radiation striking and leaving the Earth's surface.

Network

A number of electrical components connected together in a way to form a number of interrelated currents The components comprise resistors, capacitors and inductors; the network parameters are the actual values of these components.

If these values are autonomous of current, they are said to be linear, otherwise nonlinear if current will convey in both directions they are said to be bilateral, otherwise unilateral. A network without a source or sink of energy is considered passive, otherwise active.

If more than two conductors meet at a point, it is recognized as a branch point and a conducting path between two branch points is recognized as a branch. Any closed conducting loop in the networks known as a mesh.

Neutral Filter

A filter that possesses the ability to absorbs radiation of all wavelengths equally, i.e., it lessens the radiation intensity without altering the comparative spectral distribution.

Neutralization

The provision of negative feedback in an amplifier adequate to neutralize any positive feedback and so avert oscillations.

Neutral Temperature

The temperature of the hot junction of a thermocouple, the cold junction being at 0°C, for which the thermoelectric electromotive force is a highest.

Neutrino

A constant elementary particle having zero rest mass, a spin off and a speed equal to that of light. Three sorts of neutrino exist, one in each generation; for each sort there is a corresponding antineutrino. The presence of the neutrino was originally postulated to elucidate energy conservation in beta decay.

Neutron

It is basically an elementary particle of zero charge and a rest mass of 1.67492 x 10^{-27} kg. Neutrons exist in the nuclei of all atoms excluding hydrogen. A free neutron is unbalanced, decaying to an electron and a proton. The nature of the neutron was first recognized by Chadwick.

Neutron Diffraction

A technique that is used for determining the crystal structure of a solid by using it to diffract a beam of neutrons. It is considered as an alternative technique to X ray diffraction in some investigations.

Neutron Flux Density

The product of the number of free neutrons per unit volume and their mean speed. In a nuclear reactor that are intended for production of power, the neutron flux density is in the range 10^{16} to 10^{18} per meter squared per second.

Neutron Number

It is denoted by symbol N. The number of neutrons exist in the nucleus of an atom. It is the distinction of the atomic number and the mass number.

Neutron Star

A star of less than 20 kilometers diameter formed after its nuclear energy sources have been drained and gravitational collapse of the matter in the star has happened. When the density surpasses 107 kilogram per cubic meter, electrons and protons may fuse together to form neutrons. For stars not beyond a mass twice that of the Sun, strong repulsive forces between the neutrons stop the contraction and a steady neutron is formed.

Newton

It is denoted by Symbol N. The SI unit of force, which is equivalent to the force that gives a mass of 1 kilograms an acceleration of 1 meter per second per second.

Newtonian Force

A force whose magnitude is inversely proportionate to the square of the separation of the points between which the force acts.

Nicol Prism

A prism being made by cutting two pieces of calcite in specific directions and cementing them together with Canada balsam. The incident unpolarized light undergoes double refraction in the first component. The extraordinary ray conveys through the Canada balsam and arises from the second component. The ordinary ray is completely internally reflected at the Canada balsam and so the rays arc dctached. The way of polarization of the extraordinary ray is with vibrations analogous to the short diagonal of the rhomb-shaped section observed end on. Such prisms are therefore used for producing and analyzing plane polarized light.

Nife Cell

A type of accumulator that are comprising of a positive nickel and a negative iron plate dipping into sodium hydroxide solution.

Night Blindness

A condition that is allied with a shortage of rhodopsin in the retinal rods of the eye. It is thought that vitamin A deficiency obstructs the potential to resynthesize rhodopsin that has been broken down by the action of light.

Node

- A point of zero displacement of a standing wave. The parting between adjacent nodes is half a wavelength.
- A point at which the value of wave function is zero in an atomic orbital.

- A terminal of any branch of an electric network or a mutual terminal of two or more than two branches.
- The meet of the orbit of a celestial body with the ecliptic.

Noise

Noise can be defined as any sound that is unwanted or undesired.

In other words, it is also said as unwanted disturbance in an electric or communication system. If the disturbance includes a wide frequency spectrum it is recognized as white noise and is mainly comprised of random noise ascending from transient disturbances, and thermal noise.

White noise makes loudspeaker hiss and television screen snow. If the noise is because of a single momentary disturbance, or to several such at intervals, it is renowned as impulse noise. Impulse noise is accountable for loudspeaker clicks.

Nonconservative Force

A force such that the work done by it in returning an object to its initial point by a closed path is not zero. The force pertained in pushing an object round a closed path on a rough table can be referred as an instance of a nonconservative force.

Non-Euclidean Geometry

Any logical self-consistent mathematical system associated with shape but not centered on the parallel postulate of Euclid. Non-Euclidean geometry is basically used in Relativity theory.

Noninductive Winding

It is basically a type of winding. The winding turns back on itself so that any current prompted in one half of the wiring will be equivalent but opposite to that provoked in the other half. It is due to the fact that current follows the same path but in reverse directions in the two parts of the wire.

Nonlinear Optics

The study of the effects instigated by very high intensity light beams. An instance is the surge in the refractive index of glass in the path of an intense laser beam; as a result, the beam is constricted and so its intensity and that of its related electric field surge, sometimes to such a value that the glass is crushed.

Nonpolar

It usually refers to a substance that lacks any permanent dipole moment. Polar compounds will not dissolve in nonpolar solvents, but such solvents willingly dissolve other nonpolar substances.

Nonradioactive Tracers

Nonradioactive substances, often unusual isotopes, whose atoms can be recognized often with the assistance of a mass spectrograph. For instance, the average lifetime of red blood cells has been determined as about 4 month using the nonradioactive nitrogen isotope 15N: over three days subjects were fed glycine which had been created using ISN which was then taken up by the hemoglobin. Blood samples were then analyzed for I5N at regular intervals until the quantity of it returned to the normal value of 0.36%.

Normalization

The institution of a numerical factor into the equation $y = j(x)$ so that the area, if finite, under the consequent graph becomes accord. The process plays an imperative role in statistics and quantum mechanics.

Nova

A close binary star in which an explosion arises (on a white dwarf companion), creating a luminosity increase of up to 100,000 times. The luminosity then gradually declines to its original value.

Nuclear Isomers

Nuclei of the same atomic number and mass number but with distinctive radioactive properties. Isomers exhibit distinctive energy states of the nucleus.

Nuclear Magnetic Resonance

A method of investigating the spins of atomic nuclei. A sample, generally solid or liquid, is subjected to a strong variable magnetic field of magnitude about 2 tesla and also to a radio frequency field of 1–100 megahertz at right angles to the magnetic field.

Since a nucleus with spin has an allied magnetic moment, the external magnetic field yields a splitting of energy levels in such nuclei; the variance in the energy levels be contingent on the magnitude of the magnetic field strength.

For certain values of the field, the energy level variance equals the radio frequency energy and so the latter is sturdily absorbed by the sample. A signal is thus produced in the small detector coil surrounding the sample. The nuclear magnetic resonance spectrum is the plot of detector signal against magnitude of magnetic field strength.

Nuclear Medicine

The study of the use of radioisotopes in the diagnosis of disease, particularly cancer. A radioisotope behaves in the body in the similar manner as its nonradioactive counterpart and will, in specific, accrue in the same regions. An abnormally high level of radioactivity in any of these regions suggests the occurrence of cancer cells in them.

The radioactivity can be measured in a number of ways, for instance using a gamma camera, scintillation counter. The thyroid, liver, and kidneys are the principal subjects of investigation by the method.

Nuclear Reactor

An assembly in which a self-sustaining neutron chain reaction due to fission can be controlled and maintained. Various types of fuel, coolant, moderator, and cladding have been developed in the present scenario. The best choice for a specific objective is a subject of debate.

Nuclear Reactor Types

Five main categorizations exist: thermal, fast, power, breeder, and research. A power reactor is intended for maximum power output whereas the chief objective of a research reactor is experimental. In the thermal reactor the highest neutron energy is about 0.1 electron-volt. In a fast reactor the neutron energy can be range from 10 kilo-electron-volt to 2 mega-electron-volt.

A fast breeder reactor is intended at producing more fissile material than it consumes: the core is surrounded by a blanket of natural uranium in which some neutrons from the fast reactor cause the making of plutonium; the plutonium can be used to augment the fuel. Most present reactors are thermal, the type pondered to be easiest and safest to control; they do not though function as breeder reactors.

Nuclear Weapon

A weapon in which an explosion resultant from uncontrolled nuclear fusion or fission. The first weapon was the fission bomb, also known as the atom bomb or A bomb. In this device, two masses of uranium–235, each less than the critical mass, were brought jointly to form a mass greater in comparison to the critical mass. The resulting chain reaction had an explosive effect of 20 kilotons of TNT, though the uranium mass was only a few kilograms.

Nucleus

The most massive part of an atom, carrying a positive charge of Ze where Z is the atomic number of the element and e the magnitude of the charge on an electron.

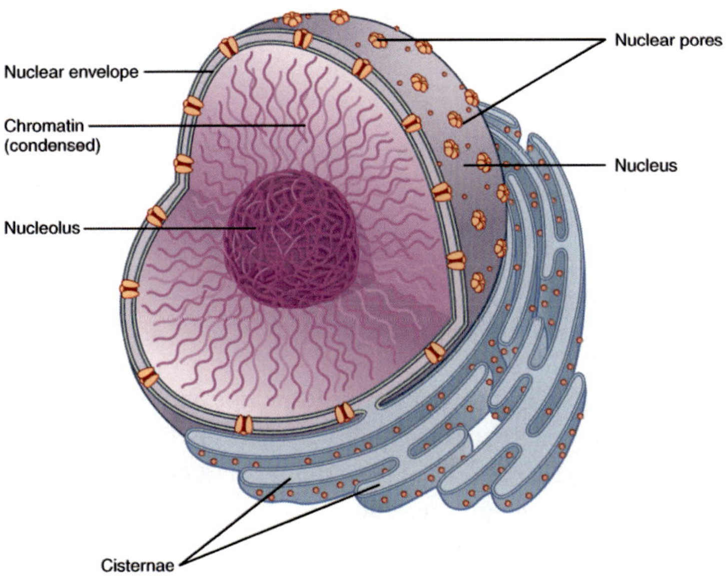

Figure 41. Nucleus.

Source: Image by Wikimedia Commons.

For nuclei of the identical element Z is static, but the mass number, A, may vary within limits and henceforth so may the number of neutrons in nuclei of the same element. A given nucleus is signified by its chemical symbol with A and Z as superscript and subscript correspondingly; thus ~He and ~He represent isotopes of helium.

Nuclide

- A nucleus is basically portrayed by its neutron number and its atomic number, i.e., nuclide refers only to a specific nuclear species.
- The atom to which a stated nucleus belongs.

Null Method

A method in which a zero reading is attained by balancing one quantity against another. The method is used for instance in the Wheatstone bridge.

Numerator

The number placed above the line in a common fraction; thus for 7/8, the numerator is 7.

Number of Poles

The number of distinctive electrical conducting paths that a switching device opens or closes concurrently.

Numerical Aperture

It is denoted by symbol NA.

The quantity n sin i, where n is the refractive index of the medium in which light strikes a microscope objective and i is half the angle subtended by the objective at the object. The larger the numerical aperture is, the better the resolving power of the microscope.

Nusselt Number

A dimensionless quantity that is defined as $hl/(\lambda\theta)$ where h, is rate of loss of heat per unit area of a hot body engrossed in material medium, the temperature distinction between medium and body being 9; I is a typical dimension of the body and λ is the thermal conductivity of the medium.

Nutation

The oscillation of the axis of a rotating body about its mean position. The phenomenon is demonstrated by a spinning top and by the Earth.

Nyquist Noise Theorem

A theorem stating where dP/df = kT, where dP/df is the rate of change of power P, because of the thermal noise, with frequency 'f' at non extreme temperature T; k is the Boltzmann constant.

O

Object

An object can be defined as anything that can be viewed using an optical system. An object is said to be real if light truly comes from it and virtual if light only seems to come from it. Thus, if rays from one optical system are interrupted by a second optical system before they have come to a attention, then the real image from the first system becomes the virtual object for the second system.

Occlusion

The duping of small quantities of one substance in another. Small amounts of solution are often blocked-in crystals created from the solution. Gas molecules or atoms may be occluded in interstitial positions in a solid.

Occultation

The withdrawal of one celestial body behind another while the first body is being witnessed from Earth.

Octahedron

A polyhedron having eight faces. If the faces are correspondent equilateral triangles, the octahedron is believed to be regular.

Octal Notation

It is basically the system that represent numbers using the base 8, i.e., integers 0–7 are employed. Thus, the number signified by 175 in octal notation is $1 \times 8^2 + 7 \times 8^1 + 5 \times 8^0 = 125$ in decimal notation. Each digit in an octal number can be signified by a three – digit number in binary arithmetic. For instance, binary 101 is $1 \times 2^2 + 0 \times 2^1 + 1 \times 2^0 = 5$. Octal is thus used in computers, each digit corresponding to one byte.

Octant

A segment of a circle subtending an angle of 45° at the center of the circle, i.e., it is an eighth of the circle.

Octave

A frequency interval between two notes in such a manner that the frequency ratio of the notes is 2:1. An octave covers 8 notes of a scale in music.

Octet

A group of eight electrons in the shell of an atom. This arrangement is extremely consistent and is prevalent in inert gas atoms except helium.

Oersted

It is denoted by symbol Oe. An electromagnetic unit of magnetic field strength in CGS units. It is the field strength that yields a force of magnitude one dyne or unit magnetic pole.

Ohm

It is denoted by Symbol 0. The SI unit of electric resistance equivalent to the resistance between two points of a conductor (that is not a source of electromotive force) whose possible difference is one volt and between which flows a current of one ampere. This ohm swapped the international ohm that was defined as the resistance at 0°C of a column of mercury of uniform cross section and length 106.3 centimeter and mass 14.4521 gram.

Ohm's Law

At persistent temperature, the current flowing through a conductor is unswervingly proportional to the potential difference applied across it. The continuous of proportionality is the resistance of the conductor.

Oil Immersion Objective

An objective used in a microscope in order to enhance the numerical aperture. A few drops of cedar wood oil are positioned on the upper slide comprising the object and the objective lowered so that its lower surface is engrossed in the oil. By suitable choice of object distance, spherical aberration due to the aim can be curtailed.

Olber's Paradox

A paradox concerned with the darkness of the night sky that should, for an infinite static and unchanging universe, be uniformly bright. Though, since these assumptions on the nature of the universe are inaccurate, the paradox is resolved.

Operating Point

The point on the characteristic curve for a valve or transistor demonstrating the current and voltage for the operating conditions being used.

Operational Amplifier

A high gain, high input impedance direct coupled amplifier generally supplied as a complete packaged unit. Such amplifiers have wide control applications and instrumentation.

Operator

A mathematical symbol demonstrating a particular operation. For instance, the differential operator d/dt represents differentiation in relation to the variable t.

Ophthalmoscope

A portable instrument for seeing the retina of the eye; A battery operated light source B together with a lens L whose location can be altered are contained in a handle backing the plane mirror M, through a hole at whose center observations are made.

Figure 42. Ophthalmoscope.

Source: Image by Wikipedia.

The subject's retina R is enlightened by an unfocused diverging light beam. Light scattered from the retina of a normal relaxed eye arises from the eye as a parallel beam, that is focused by E, the observer's relaxed eye. To accurate for any deviations from normal refraction by the subject's eye, provision is made for initiating a suitable correcting lens behind the hole in the mirror; a clear view of the subject's retina can therefore always be attained.

Opposition

A situation arising when a celestial body lies on the line, not formed, joining two other celestial bodies. If the body lies on the formed line, conjunction is said to arise. It is applicable for an outer planet such as Mars or Jupiter. At full moon the Moon and Sun are in opposition in relation to Earth; at new moon they are in conjunction.

Optical Activity

The property of some crystals, solutions, liquids, and vapors of rotating the plane of polarization of plane polarized light. The amount of rotation is in proportion to the length of substance crossed and, where suitable, to the concentration; it also contingent on the wavelength of the polarized light and on the substance temperature. The rotation may be either laevorotatory or dexfrorotatory. Optical activity is measured in a polarimeter

Optical Axis

- The line joining the centers of the entrance and exit pupils of a centered optical system. The cardinal points lie on it.
- The direction, in a crystal showing double refraction, in which the extraordinary and ordinary rays are not separated, i.e., in which they travel with the similar speed.

Optical Bench

A track on which optical components can be affixed and moved. It is used widely in optics experiments.

Optical Center

The axial point through which a ray, guided before entering an optical system towards a nodal point, really conveys. For thin lenses the optical center coincides with the geometric center of the lens.

Optical Character Recognition

A technique of feeding conventional printed characters into a computer. A special type face is generally imperative.

Optical Flat

A surface which nowhere diverges from flatness by an amount larger than the wavelength of light. Such surfaces are imperative for many optical devices and experiments and are tested by interference methods.

Optical Isomerism

The phenomenon shown by substances whose physical properties, apart from optical activity, are identical. The phenomenon is extremely common in substances that comprise asymmetric carbon atoms, for instance the two forms of lactic acid, for which the molecular structure of one isomer is the mirror image of the other.

Optical Path

The product of the actual distance crisscrossed by a light ray and the refractive index of the medium in which it is travelling. If the ray voyages through various media, the total optical path is the sum of the optical paths in the several media.

Optic Nerve

The nerve bundle that is accountable for transmitting electric signals produced in the cones and rod of the retina to the brain.

Optics

A branch of physics that is concerned with the study of the production, measurement, propagation, and properties of light. The ray treatment of light is known by the name geometric optics and the wave treatment is referred as physical optics.

Optics Sign Conventions

The two main conventions are 'real is positive' and the Cartesian sign convention. For the former, any distances measured from an optical component are contemplated positive if they end on a real entity and negative if they end on a virtual one.

In the Cartesian convention, light is drawn moving from left to right and positive distances are those computed from an optical component to an entity in the track of the incident light; negative distances are those assessed from an optical component beside the direction of the incident light; transverse distances above the optic axis are positive and those below of it are negative.

Optoelectronics

- It refers to the study of an electric field on optical phenomena.
- The technology that is intended at the production of optical signals from electrical signals and the detection of such signals, used for instance in transmission of signals along optical fibers.

Orbit

- The path, often an ellipse, of one celestial body around another.
- The path, often envisaged as an ellipse or circle, of an electron around the nucleus of an atom.

Order-Disorder Transformation

- The change in a solid solution from a state in which atoms of the several components take up favored sites in the structure (i.e., an ordered phase) to a state in which the atoms are dispersed at random over these sites (i.e., a phase of disorder).
- The reduction in the static dielectric constant of a solid dielectric with rising temperature, analogous for dipolar materials to a variation from a state of orientation order to one of disorder.

Order of Diffraction

A whole number that exemplifies a diffraction spectrum by its displacement from the beam that is undeviated. Overall, for each order, there are two spectra, one on either side of the undeviated beam.

Order of Interference

An integer that typifies an interference fringe by the number of wavelengths in the path differentiation between the beams giving rise to it.

Order of Magnitude

An estimated magnitude to within a factor of 10; thus 0.01 and 0.04 are of the same order of magnitude but 10 and 1000 differ by 2 orders of magnitude.

Oscillation

A periodic energy difference in an electrical, atomic or mechanical system. An electric oscillation may be confirmed by charging a capacitor connected in corresponding to an inductor. The charged capacitor discharges through the inductor that initiates to store magnetic energy at the expense of the electric energy of the capacitor and endures to do so until the capacitor is discharged fully. The capacitor then commences to recharge but in the opposite direction until all the energy is in the electric field once more. The current then converses and the whole process arises in reverse order and then starts again so that continuous electric oscillation results, as specified.

Osmosis

The favored transmission by a semipermeable membrane of certain substances in solution. For instance, an aqueous solution of sugar in a thistle funnel wrapped by a piece of parchment and placed in water, adds water and preserves the sugar; parchment thus lets the passage of water molecules but not sugar molecules.

Osmotic Pressure

It is denoted by symbol IL. The minimum pressure that can be applied to a solution, averts osmosis. For dilute solutions of nonelectrolytes, ITV= RT where Vis the volume of solution comprising unit mass of solute, R the universal gas constant and T the absolute. An electrolyte solution shows a greater osmotic pressure in comparison to the equation predicts since electrolytes ionize in solution.

Otto Cycle

The thermodynamic cycle, upon which the operation of spark ignition engines is dependent. PQ resembles to adiabatic compression of an explosive air-petrol mixture by a piston. Q is the firing point and along QR the temperature surges greatly at constant volume. RS corresponds to adiabatic expansion; the piston being pushed back. In addition to the SP, the temperature falls at persistent volume; as the piston returns, the exhaust gas is swept out and a fresh petrol-air mixture taken in to resume the cycle.

Output

- The signal, voltage, current, etc. delivered by an electric device or circuit.
- The terminals or other place where the signal is being conveyed.
- The processed data carried by a computer.
- The part of the computer system that transforms the data into usable form. Instances are the printer, magnetic tape, the punched card producer, and visual display unit.

Oxygen

It is denoted by symbol 0. An element forming 28% of the atmosphere by volume, formed by liquefying air. Oxygen is imperative for combustion and for the respiration of animals and plants. It has wide industrial as well as medical uses.

Oxygen Point

The temperature of equilibrium between gaseous and liquid oxygen at a pressure of 1.01325×10^5 pascal. The temperature, 90.188 K, is taken as a fixed point in the international temperature scale.

Ozonosphere

A layer in the Earth's atmosphere comprising ozone. It extends from 15 kilometers to 30 kilometers above the Earth's surface and absorbs the Sun's higher-energy ultraviolet radiation.

P

Packing Fraction

A quantity stated as $(A_r–A)/A$ where A_r is the relative atomic mass of an isotope and A its mass number. The value for ^{16}O is zero, becoming negative for higher values of A up to A = 180 and attaining minimum for A = 50. Positive values designate uncertainty and so substances with A values in the range 0–16 are used in nuclear FUSION processes and those with A values beyond than 180 in nuclear fission processes.

Pair Production

The production of an elementary particle and its antiparticle from a photon entering the field of an atomic nucleus. Thus, a gamma ray photon of energy higher than 1.02 megaelectron volt, passing near to an atomic nucleus, may be transformed into a positron and electron.

Paleomagnetism

The study of the residual magnetization of rocks in order to clarify the past behavior of the Earth's magnetic field. Iron compound bearing rocks, created at high temperatures, were attracted by the Earth's field while they cooled and so the direction and intensity of the residual magnetism yields information about the direction and intensity of the Earth's field at the time of the rocks' formation; this time can be found by radioactive dating. Such studies display that the Earth's polarity has retreated, i.e., south and north poles have interchanged, many times in the past.

Parabola

A conic of eccentricity equivalent to 1. The equation of the parabola is y2 = 4ax having its emphasis at the point (a, 0); its directrix is the line x = –a. The vertex is at the beginning and the axis coincides with the x axis. A line drawn from the emphasis to any point on the parabola makes the same angle with the normal to the parabola at that point as does a line drawn from the point corresponding to the x axis: a property of importance for the parabolic mirror.

Paraboloid of Revolution

The shape attained by rotating a parabola about its axis. Thus, cross sections perpendicular to the axis are circles and cross sections in planes parallel to the axis are parabolas. Rays of light incident equivalent to the axis on a reflecting surface of such a shape will all pass through the focus, and a source placed at the emphasis will give augmentation to an precisely parallel beam of light. Reflecting paraboloids of revolution thus find radio and optical telescope application and are also used in electric fires, search lights, etc.

Parallax

The apparent change in the parting between two objects when they are observed from distinctive positions. Objects in line with the eye seem to move apart as the observer moves to right or left, unless the objects coincide. This is the foundation of the no parallax test for finding an image position in elementary optics. Parallax can arise in measuring instruments when the pointer is a small distance in front of the scale. Henceforth, in many instruments a mirror is positioned behind the pointer, whose position is read when it is in line with its image, thus abolishing the error.

Parallel

- Denoting planes and lines that are the same distance from each other at all points.
- Denoting a mode of connection of electric apparatus such that the current splits between them.

Parallel Axes Theorem

The moment of inertia of a body about any axis is equivalent to the sum of its moment of inertia about a corresponding axis through the center of mass and the product of the mass of the body and the square of the separation of the axes.

Parallelogram

A quadrilateral that has both pairs of opposite sides parallel. Its area is the product of the length of a side and the perpendicular distance of that side from its opposite side.

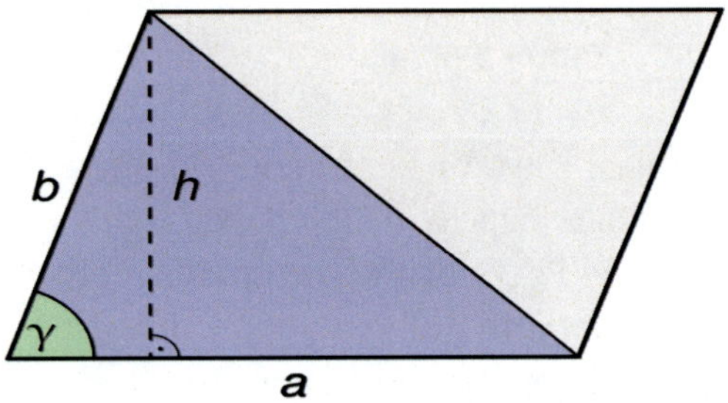

Figure 43. Parallelogram.

Source: Image by Wikimedia Commons.

Parallelogram Rule

If a vector is entirely exemplified in direction and magnitude by a side of a parallelogram and another vector is likewise represented by an adjacent side, then the subsequent of the vectors is signified in magnitude and direction by the parallelogram diagonal sketched from the meet of the adjacent sides. Common instances of the application of the rule are to velocities and forces.

Paramagnetism

The phenomenon displayed by those substances which, under the guidance of an applied magnetic field become weakly magnetized in the course of the field but lose this directional magnetization when the field is detached.

A paramagnetic substance is deemed as an assembly of magnetic dipoles that are normally engaged at random because of thermal agitation. For small fields and high temperatures, the magnetization produced is proportional to the field strength; at high field strength or low temperature, saturation is attempted.

The dipoles are thought to ascend from electron magnetic moment because of orbital and spin motion in molecules of atoms comprising unpaired electrons.

Parameter

A term in a mathematical expression; each value of the term is related with several sets of values of the other variables. For instance, the expression x2 + y2 = r2 describes a circle radius r center the origin. Each distinctive value of the parameter r corresponds to a specific circle so that the expression defines a family of concentric circles.

Parasitic Oscillations

Unwanted oscillations which may arise in a circuit. Parasitic oscillations are generally of much higher frequency in comparison to those for which the circuit was intended.

Parity

It is denoted by symbol P. A symmetry property of a wave function. If the signs of all the coordinates in a wave function are altered and the value of the wave function remains the same, the parity is said to be + 1; however, if the value of the wave function changes consistently after the substitution, then the parity is said to be –1. The parity of the total wave function defining a system of elementary particles is preserved in electromagnetic interaction and in strong interaction but not in weak interaction: for instance, in beta decay the emitted electrons always spin in the contradictory sense to the direction of motion and so parity is not preserved.

Parsec

It is denoted by Symbol pc. A unit of length used in astronomy, equivalent to the distance corresponding to an annual parallax of 1 second of arc, i.e., to 3.086×10^{16} meter of 3.26 light year.

Partial

Any pure tone component of an intricate tone, the frequency being an integral multiple of that of the fundamental, which is not a partial. Some musicians though use harmonic and partial as synonyms.

Partial Derivative

A derivative in relation to one variable only, and other variables existing being treated as constants. The symbol o is used. Thus, if $y = 3x^2z^3$ then $dy/dx = 6xz^3$ and $dy/dz = 9x^2z^2$.

Particle Physics

The branch. of physics that is interested with the properties, interactions and structure of elementary particles.

Particle Velocity

It is denoted by symbol u. The alternating component of the velocity of a medium transmitting sound, i.e., the total velocity of the medium less the velocity not because of sound propagation.

Particle-Wave Duality

A phenomenon that is applicable to all physicality's. The choice between wave description and particle description is wholly a matter of convenience. The particle and wave aspects are linked through the relations $E = hv$ and $p = h/\lambda$), where the energy E and momentum p refer to a particle while the frequency v and wavelength λ denote to a wave. The Planck constant h appears in both equations in relation to the relativity theory.

Parton

An ultimate fundamental particle proposed as a basic unit of other fundamental particles. In the simplest theory the patron is a quark.

Pascal

It is denoted by symbol Pa. The SI unit of pressure, equivalent to a pressure of 1 newton per square meter.

Pascal's Law

Whenever an external pressure is applicable to any confined fluid at rest, the pressure is augmented at every stage in the fluid by the quantity of the external pressure.

Paschen-Back Effect

An effect that is akin to the Zeeman effect but concerned with magnetic fields adequately strong for both the electron orbital and electron spam angular momentum vectors to each distinctly take up their possible orientations relative to the field direction. The consequent hyperfine structure produced in the spectral lines varies from that of the Zeeman effect.

Paschen Series

A series in the spectrum of the hydrogen atom. It can be stated by the equation $1/\lambda = R(1/3^2 - 1/n^2)$ where A. is the wavelength of a line, n is any integer larger than or equal to 4 and R is the Rydberg constant. The lines are in the infrared part of the spectrum.

Paschen's Law

The breakdown voltage for a discharge between electrodes in gases is a function of the product of electrode separation and gas pressure.

Passive Component

An electronic component that is incapable of an intensifying function. Instances include are resistors, inductors and capacitors.

Patching

A technique that has applicability in an analog computer to let temporary connection of circuits.

Pauli Exclusion Principle

The principle that no two electrons in an atom can have the identical set of quantum numbers. Therefore, if two electrons in an atom have the similar n, I and m values, i.e., share an atomic orbital, then they must have opposite spins since only the values ±V2 are accessible for the spin quantum numbers. Two such electrons are believed to be spin paired.

Pauli Spin Matrices

It is basically a set of matrices initiated by Pauli in relation with electron spin in nonrelativistic wave mechanics.

Peltier Effect

A thermoelectric effect in which route of an electric current through a junction between two distinctive solids causes heat to be absorbed or created at the junction, in reference to the direction of the current. Thus, semiconductor junctions can be used as cooling elements.

P-E Model of Nucleus

A model that is dependent on the assumption that nuclei are composed of electrons and protons. It failed to account for various experimental observations and was thus abandoned.

Pencil of Rays

A slender cylinder or cone of rays traversing an optical system and restricted by a STOP.

Pendulum

A body deferred so as to be free to oscillate. Pendulums have been used in several determinations of the acceleration because of gravity. A simple pendulum can be defined as a small mass that is suspended from a point by a thin thread and swings in a vertical plane. For small oscillations its period of oscillation is given by $2\pi(l/g)^{1/2}$, where 1 is the thread length and g the acceleration due to gravity.

Figure 44. Pendulum.

Source: Image by Wikimedia Commons.

A conical pendulum is basically a simple pendulum whose bob swings in a horizontal circle. For circles of small radius, the period is identical as that of the simple pendulum of the same length.

Pentode

A thermionic valve with five electrodes. It is a TETRODE modified by the addition between screen grid and anode of a third grid that is given a negative potential in relation to both screen grid and anode. Low-velocity secondary electrons from the anode are thus barred from reaching the screen grid.

Percolation Limit

It is basically concentration of magnetic element, in a disordered crystalline alloy having one constituent with a magnetic moment, above which the ferromagnetic state swaps the state of random orientation of atomic spins.

Period

The constant interval between identical states of a system having properties that vary periodically. For a system in which a quantity differs with time, the period is the time taken to complete a cycle; thus, for a body moving round a circle with angular velocity w the period is $2\pi/w$. The period of a crystal lattice in a specific direction is the parting of similar lattice points along that direction.

Periodic Table

A classification of chemical elements with reference to atomic number. The sequence of elements breaks up into seven eight vertical groups and horizontal periods. The elements in each group have alike electron configurations and thus show marked similarities in behavior.

Periscope

An apparatus for observing objects when there is no direct line of sight to the eye. In its simplest form, the instrument comprises two parallel mirrors at 45 degrees to the direction of view. Light from the object is twisted through 90 degrees by the top mirror, strikes the lower mirror and is again turned through 90 degrees to enter the eye of observer. Total internal reflection by prisms, in comparison to mirror reflection, is used in better quality instruments such as submarine periscopes.

Permanent Gas

A gas that cannot be liquefied by pressure alone at usual temperatures, i.e., a gas needing cooling to reach its critical temperature.

Permanent Magnet

A magnetized mass of ferromagnetic substance, steady against sensible handling and of high remanence. A definite demagnetizing field is needed to eradicate the magnetism.

Permeability

- absolute permeability. It is denoted by Symbol μ. The ratio of magnetic induction to the external magnetic field strength persuading it. The unit is henry per meter.
- relative permeability. It is denoted by symbol μ_r. The ratio of the absolute permeability for a given medium to the absolute permeability of a vacuum.

Permittivity

- Absolute permittivity. It is denoted by symbol E. The ratio of electric displacement to the electric field strength yielding it. The unit is farad per meter.
- Relative permittivity. It is denoted by symbol Er. The ratio of the absolute permittivity for a given medium to the absolute permittivity of a vacuum. The relative permittivity of the dielectric medium of a capacitor can also be stated as the dielectric constant.

Perrin Tube

A spherical evacuated glass bulb with a side arm. A magnet placed will deflect the electron beam at R from hot cathode C into the side arm P, instigating the negatively charged leaf of the gold leaf electroscope E to ascend even further and thus representing directly the negative charge on the electron.

Perpetual Motion

Perpetual motion is defined as a continuous motion without any supply of energy. This type of motion would only be possible in the absence of any kind of frictional forces. The production of the useful work with the help of a perpetual motion machine would disrupt the first law of thermodynamics.

Persistence

Persistence is the interval of time after excitation ceases in which the light is being emitted from the screen of a cathode ray tube. Certain gases show weak luminosity for a good amount of time after the cessation of an electric discharge in them.

Phase

Phase is the state of vibration of a system which is varying periodically at a given point of time. Systems that are vibrating with the same frequency are said to be in a phase if their maximum and minimum values that took place at the same time; if not, then they will be said as out of phase.

In expressions $y = A \sin (nt)$, $y = A \sin (nt + B)$, for two sinusoidal vibrations, B represents the phase difference between them.

Phase Modulation

Phase modulation is the modulation in which the carrier wave phase is varied about its unmodulated value by an amount proportional to the signal amplitude and at the frequency of the signal, the carrier wave amplitude remaining constant.

Phase Plate

Phase plate is a transparent plate carrying an annular groove, and it is placed in such a way that light passing through the grove varies by a quarter of a wavelength in optical path from light passing through the entire plate.

Phon

Phon is a unit of loudness of sound, and it is defined with respect to a standard tone of frequency that is 1000 hertz. The intensity of the standard is differed until it is judged to have the same loudness as the sound under the investigation, and the intensity of the sound is recorded. If the measured value is 'n' decibel above the threshold of hearing, in that case, the loudness is said to be phon.

Photocell

Any light-electricity transducer is said to be a photocell. The earliest kind was a vacuum tube which is consisting of a photocathode and photo-anode.

When the light fell on the surface of the cathode, a current flowed when the anode was maintained at a positive potential with respect to the cathode.

Modern photocells are consisting of a bar or thin polycrystalline film which is made with the help of a semiconductor with suitable contacts. Illumination produces an increment in the conductivity because of the production of charge carriers; the current is recorded.

Photochromic Substance

Photochromic substance is a substance which changes color when the light falls on the surface of it. Often, when the light is cut off, the original color returns to its original state.

Photoconductivity

The increment in the electric conductivity of certain solids, generally the semiconductors like selenium, when it is exposed to electromagnetic radiation, the process is said to be photoconductivity. It took place when the photons have adequate amount of energy in order to raise electrons from a filled band to the conduction band.

Photo-Detachment

Photo-detachment is defined as the removal of an electron with the help of a photon from a negative ion in order to give a neutral atom or molecule.

Photodiode

Photodiode is a type of semiconductor diode which is sensitive to the light. When it is operated, the diode has reverse bias. Small amount of carriers flows in the circuit and create a dark current. When the junction is illuminated more hole–electron pairs are generated, and these hole–electron are then swept across the junction, constituting the light current.

Photodisintegration

Photodisintegration is the disintegration of an atomic nucleus with the help of a gamma ray photon or an X ray photon.

Photoelasticity

Photoelasticity is the study of the effects of stress in transparent materials on light traversing them. Generally, under stress isotropic transparent materials possibly show double refraction; marked effects are then produced on illuminating them with polarized light.

Photoelectric Effect

Photoelectric effect is the ejection of electrons from a solid in the form of irradiation with electromagnetic radiation of suitable frequency. The number of electrons that have been emitted relies on the intensity of the incident radiation, not its frequency. Electrons are only emitted when the frequency of the radiation exceeds a certain value, which is characteristic of the substance and is known as the photoelectric threshold.

Photoemission

Photoemission the emission of electrons with the help of a substance in the form of a result of bombardment by photons, as in the photoelectric effect and in photoionization.

Photography

Photography is defined as the production of more or less permanent images with the help of exposing a suitable material to radiation. And it is followed by the treatment of the material with the help of chemicals. Generally, the suitable material is a silver halide which is embedded in emulsion on film.

With the help of a photoelectric effect, exposure of the film causes conversion of silver ions to silver atoms, forming a latent image in the emulsion. By treatment with a mild reducing agent, known as a developer, opaque specks of metallic silver are formed round each silver atom of the latent image.

Photoionization

Photoionization is defined as the ionization of atoms or molecules with the help of electromagnetic radiation. Only a photon which is having energy greater as compared to the first ionization potential of an atom, or molecule, can remove an electron from it. The difference in photon energy and first ionization potential is shared between the ion and electron kinetic energies.

The share of the ion is negligible and the reason behind that is, it is very much more massive as compared to the electron. In this way, if E is the kinetic energy of the electron, $E = hv - I_1$.

Photoluminescence

Luminescence is an outcome from the irradiation which occurs with the help of electromagnetic radiation. Absorption of this radiation raises the atoms or molecules of the luminescent substance to excited states. As the atoms or molecules return to the ground state, either directly or via an intermediate excited state, radiation of longer wavelength than the exciting radiation is emitted. Practical use is made of the phenomenon in fluorescent paints and materials. It is also utilized in detergent whiteners, that assists in absorbing the ultraviolet radiation and then release blue light over a long time period, thus giving white fabric a blue cast and counteracting any yellowing.

Photometer

Photometer is an instrument that is being employed in Photometry. The main use of the photometer is to compare the light sources. Generally, a photometer compares an unknown with a standard one. Both visual photometers as well as physical photometers are available. For the former the eye is used to make the comparison, while for the latter a photocell, thermopile and bolometer may be employed.

Photometry

Photometry is used to measure the intensity of the light and amounts of illumination. The radiation may either be calculated as per its visual effects, that is, with the help of human eye, and that's how its energy may be determined. When the eye is used the adjective luminous precedes the physical quantity, however, for the measurement of energy, the adjective radiant precedes the quantity.

Photomultiplier

Photomultiplier is a device which is used for detecting photons. Light incident on the photocathode of the tube causes electron emission with the help of the Photoelectric Effect. These electrons are accelerated with the help of a potential difference to the first electrode where, by Secondary Emission, they produce the emission of nearly 2 to 5 electrons per incident electron. The process is being repeated at the second electrode, which is at a positive potential with respect to the first electrode, and so on.

Photosynthesis

Photosynthesis is a natural phenomenon, in which the synthesis take place due to the action of light. A very important example is the conversion with the help of plant leaves of atmospheric carbon dioxide to carbohydrates in the presence of light.

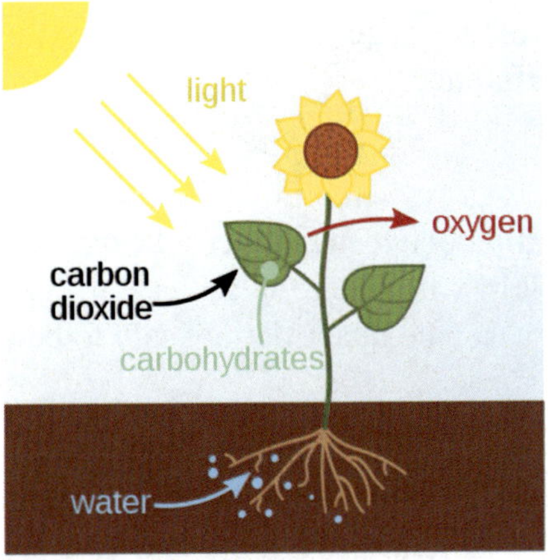

Figure 45. Photosynthesis.

Source: Image by Wikimedia Commons.

Phototransistor

Phototransistor is a bipolar junction transistor which is connected with the floating base electrode. The base signal is supplied with the help of excess carriers that is generated by the illumination of the base. The emitter current relies on the illumination until the establishment of equilibrium between the base recombination and carrier generation, when the emitter current saturates.

Q

Quadrant Electrometer

Quadrant electrometer is defined as an electrometer, in which a light metal vane is suspended inside four metal quadrants with the help of a torsion wire. Opposite pairs of quadrants are linked and the potential difference is being applied between the pairs. When the vane is given a potential large as compared to that of the quadrants, its deflection is proportional to the quadrant potential difference.

Quadrature

Quadrature is the relation between two periodic quantities which are having the identical wave form when there is a phase difference of 90° between both of them. In that case, the quantities are said to be in quadrature.

Quadripole

Quadripole is an electric network which is consisting of only four terminals. Out of four terminals, there are two input and two output terminals. If on interchange of input and output terminals the electric properties remain unchanged. In that case, the Quadripole is said to be symmetrical in nature.

Quality Factor

Quality factor is a number assigned to a system which is undergoing a forced oscillation. Quality factor is defined as 2n times the maximum energy which is being contained in the system divided by the average energy dissipated per oscillation. The quality factor is directly proportional to the Resonance. It is thus desirable for radio receivers to have a high-quality factor, as it requires a highly selective response.

Quantized

The term 'Quantized' is used to denote a physical quantity which can only take certain discrete values in a particular system. In this way, the quantity changes in steps and cannot vary continuously.

Quantum Field Theory

Quantum field theory is a theory which uses suitable or proper operators, obeying certain commutation relations, in order to represent all the physical observables in a system. The total energy, momentum, charge, etc. of the field is built up additively from the individual contributions to each of these variables of the particles present

Quantum Mechanics

Quantum mechanics is a mathematical theory which is derived with the help of quantum theory. The main concern of the quantum mechanics is with the mechanics of atomic and related systems in the context of measurable quantities. Quantum mechanics is consisting of wave mechanics and matrix mechanics. In relativistic quantum mechanics the simple theory was extended in such a way that the principle of relativity is also satisfied; the properties of electron spin then followed naturally from the relativistic form of Schrodinger's equation.

Quantum Number

Quantum number is defined as, any of several numbers which all-together characterize the state of an atomic or molecular system. The numbers are directly related to the eigen functions that are linked with the Schrodinger's equation. Elementary particles also have quantum numbers to describe their properties.

Quantum Optics

Quantum optics is a type of optics which is based on the concept that light is emitted from a source in the form of discrete packets of waves called quanta or photons. These discrete packets of waves which individually act like particles but which in large numbers can behave like waves. The subject is thus essentially mathematical. The analysis leads to conclusions in agreement with experimental results.

Quantum Statistics

Quantum statistics is a type of statistics that deals with the distribution of particles of a given type between the numerous quantized energy levels available. There are two types of quantum statistics, first is Bose-Einstein Statistics and second is, Fermi-Dirac statistics. Both, Bose-Einstein statistics as well as Fermi-Dirac statistics lead to the classical Maxwell-Boltzmann distribution law at adequately high temperatures, that is, when a large number of energy levels are excited.

Quark

Quark is a fundamental constituent of every matter. Every hadron, that is, Meson and Baryon, is well-thought-out to have a quark substructure. On the other hand, a Lepton does not have any substructure. There are various sorts of quark, starting from bottom quark, charmed quark, down quark, strange quark, top quark, and up quark. Out of all, at present interval of time, the bottom quark is the most massive quark; like the strange and down quarks, its charge (taking the proton charge as 1) is –1/3.

However, among all the quarks, the least massive quark is the up quark, which is having charge of 2/3. The up quark is the first-generation partner of the down quark, both of which are made up with the help of the proton and the neutron.

Quark Confinement

Quark Confinement is a theory which states that, there is some reason why quarks may exist inside elementary particles but not be observable in any experiment.

Quarter Wavelength Line

Quarter wavelength line is a transmission line a quarter of a wavelength long. Quarter wavelength line is used in the form of an impedance matching device. The main use of quarter wavelength line is in the higher ratio frequency systems.

Quarter Wave Plate

Quarter wave plate is a thin mica or quartz sheet, which is being cut in such a way that its interference color in white light is pale grey, and it is having a thickness in such way that a quarter wavelength path difference is introduced between the unseparated ordinary and the extraordinary rays into which normally incident unpolarized sodium light is split. When light which is plane polarized with its vibration plane at 45° to the principal plane of the plate is incident on the plate, circularly polarized light emerges; if the angle between the planes differs from 45°, elliptically polarized light emerges.

Quartz

Quartz is a natural crystalline form of silica, and it is colorless or of natural white. Quartz shows double refraction and may be laevo- or dextrorotatory, the amount of rotation fluctuating with wavelength. The transmission range of quartz is ranging from 180 nanometers to 4000 nanometers. It also depicts the Piezoelectric Effect.

Figure 46. Quartz.

Source: Image by Wikimedia commons.

Quartz Iodine Lamp

Quartz iodine lamp is made of tungsten filament electric lamp, which is molded in a quartz envelope filled with the iodine vapor. The working temperature for quartz iodine lamp is ranging between 500°C and 600°C. Electric energy is converted to light much more effectively as compared to an ordinary tungsten filament lamp.

Quartz Oscillator

Quartz oscillator is a circuit whose resonant frequency of oscillation is stabilized with the help of a Quartz crystal.

Quartz Wedge

Quartz wedge is a very small angled wedge of quartz, with the help of which any desired thickness (within the limits) of quartz can be inserted in front of the objective of a polarizing Microscope. By adjusting the wedge, it is possible to cancel out the phase difference introduced by the specimen and in order to reduce the value of this phase difference from the thickness of quartz that is being used.

Quasar

Quasar is a compact extragalactic source of very strong electromagnetic radiation. Most of the electromagnetic radiation of quasar is in the infrared. Generally, in most of the quasars, they have a very large red shift, and the reason behind that is the Doppler Effect, which would indicate an massive speed of recession and hence very great distance.

Quasi Particle

Quasi particle is a system of several various interacting particles which has properties just same as particles. However, Quasi Particle does not exist as a free particle, an example of Quasi particle is a Phonon.

Quench

Quench is a type of capacitor or resistor or combination of the two placed across contacts, for example the make and break contacts of an Induction Coil, in order to inhibit sparking when the current is cut off to an inductive circuit.

Quenching

- Quenching is a process of fast cooling of a hot metal, used for example in steel hardening.
- Quenching is also a sudden termination of discharge in a Geiger Counter
- Quenching is a process in which the lifetime of excited atoms, ions etc. is reduced with the help of treatment with a substance which deactivates them.

Quick Freezing

The term 'Quick Freezing' is used in the preparation of food stuffs for cold storage by means of passing them through a temperature region of 0°C to –4°C as quickly as possible. Maximum ice crystal formation results and so damage to plant and animal tissues, and this happened because of the formation of large ice crystals, is avoided.

Quincke's Tube

Quincke's Tube is a device which is utilized in order to demonstrate the interference of sound.

R

Radar

Radar is a system which is used for detection of the objects that are placed at distance with the help of reflecting electromagnetic waves off them. The time taken by the waves to travel to the object and back allows the distance of the object to be calculated.

The direction of the object is given by that of the receiving aerial. The waves can be transmitted both ways continuously as well as in pulses. The transmitting and receiving aerials can be rotated to scan any sort of area. The return signals are fed to the cathode ray tube, where the signals may be displayed in a variety of ways, it depends on the type of scanning.

Radiation Damage

Radiation damage is defined as, any harmful change that is caused by the exposure to the radiation. The damage may result from transmutation, electronic excitation, displacement the of atoms, or ionization. In living material, these sort of processes can change the genetic structure of cells, interfere with their division or kill them, leading to long-term damage to the whole organism.

Radiation Temperature

Radiation temperature is the surface temperature of a celestial body. The Radiation temperature can be calculated with the help of Stefan's Law, which relies on the assumption that the body behaves as a Black Body.

Radiation Units

Radiation units are the units that are used to express the activity of a Radionuclide and ionizing radiation dose. There are different recommended SI units are the becquerel, symbol Bq, the gray, symbol Gy, the sievert, symbol Sv, and the coulomb per kilogram, respectively replacing the CURIE, RAD, REM and RONTGEN. The temporary use of these old units is approved in order to allow time for familiarization with the new ones.

Radio

Radio is system, in which the communication of signals is done with the help of Radio Frequency Radiation. All forms of these types of communication, such as television, are included in the term, which is however sometimes used particularly for sound broadcasting.

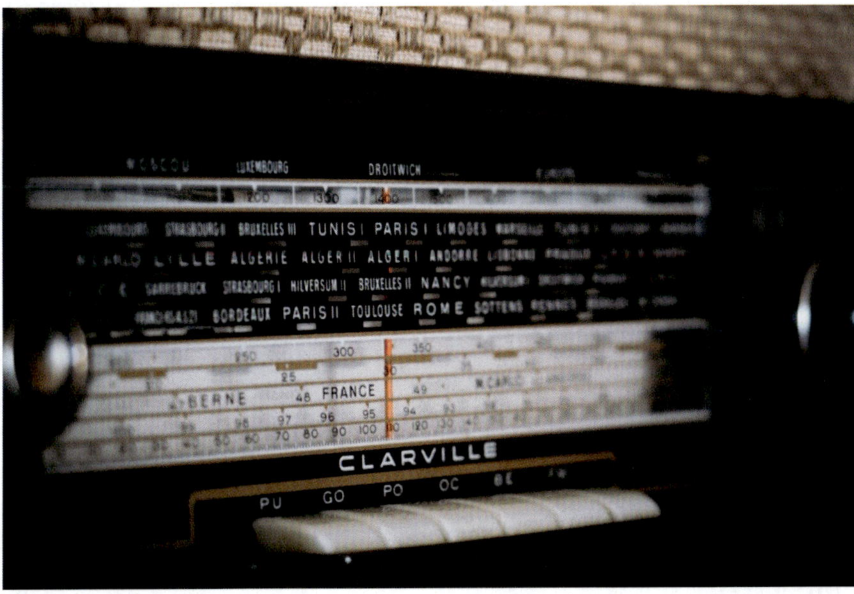

Figure 47. Radio.

Source: Image by Wikimedia Commons.

A radio transmitter produces a Carrier Wave which is of fixed frequency, and it is done with the help of modulation of this wave that the information is transmitted. The modulated wave is sent from a transmitter aerial and carried either as a ground wave or a sky wave to a receiver aerial where it is being demodulated.

Radioactive Series

Radioactive series is a series that is made of radioisotopes. And it is made in such a way that each member of the series is formed by the decay of the preceding member. There is total four such series exist, and they are as follow:

- the uranium,
- thorium,
- actinium, and
- neptunium series.

The parents of these series are respectively uranium–238, thorium–232, uranium–235 and plutonium–241.

Radioactive Tracer

Radioactive tracer is a definite quantity of radioisotope which introduced into a mechanical or biological system in order to enable its route with the help of the system and its regions of concentration to be monitored. This is done with the help of a radiation detector placed outside the system.

Radioactive Waste

Radioactive waste is defined as, any waste material which is consisting of radionuclides. Radioactive waste may produce from the mining of radioactive ores, from the manufacture of nuclear weapons, from nuclear power stations, and from various laboratories.

Radioactivity

Radioactivity is the spontaneous disintegration of certain types of atomic nuclei with the emission of Alpha Particle, Beta Particle or Gamma Ray. If the nuclei are naturally taking place, the process is called as natural radioactivity. If the radioactivity is induced with the help of irradiation, for instance, neutrons, the process is known as artificial radioactivity. Both natural radioactivity as well as the artificial radioactivity decrease exponentially with time.

Radio Astronomy

Radio astronomy is the branch of astronomy that deals with the radio frequency emissions of celestial bodies with the help of Radio Telescopes.

Radiocardiography

Radiocardiography is a technique, in which barium–137 of activity 10 millicurie is injected into the subclavian vein quickly, in such a way that it enters the right ventricle of the heart almost immediately. The flow is observed with the help of suitably placed radiation counters.

Deviation of the Radioactivity from the standard pattern allows quick diagnosis of the heart defect. The alternative technique of working a catheter from the bloodstream to the heart is much riskier and it takes a lot of time.

Radiolucent

Radiolucent is a term which is used to denote materials that are partially transmit ionizing radiation. However, the materials that are passing most of the incident radiation are described as radiotransparent and, on the other hand, the materials those passing almost none of it are described as radio-opaque.

Radio Telescope

Radio telescope is a type of instrument which is used for detecting radio waves from a particular direction in space. The simplest instrument is consisting of a steerable paraboloid dish aerial and ancillary amplifiers.

In order to that the dish may reflect the incoming radio waves to its focal point, it is essential for its surface to be accurately constructed; hence, for incoming radiation of 1 centimeter wavelength, a 100 meters diameter dish must be accurate to the nearest millimeter.

Radiotherapy

Radiotherapy is a therapy, which used for ionizing the radiation for medical treatment. In radiotherapy, the most commonly used radiation is X Rays. A beam of Gamma Rays is also frequently used and is generally attained with the help of cobalt–60.

Electron, proton and neutron beams are not so much used but Alpha Rays are being used couple of times. The cross-fire technique is used frequently: a deep – seated region is irradiated from various directions, to lessen the damage to healthy tissue. On the other hand, the radiation beam may be moved around the region under irradiation.

Reactance

Reactance is represented by the symbol X. Reactance is a property of an alternating current circuit which is consisting of an inductance and/or capacitance. For a purely inductive circuit the reactance is 2πfl., nevertheless, for a purely capacitative circuit, the reactance is 1/(2πfC).

In which, f is the current frequency, and L and C respectively represent inductance and capacitance.

Red Shift

Red shift is the displacement of the wavelengths of lines in the spectrum of a moving body in the direction of longer wavelengths. The main reason of this shift is the Doppler Effect and shows that distant stars and galaxies which demonstrate the effect are receding from the Earth, the amount of shift increasing with the speed of recession.

In the context of the objects that are beyond the galaxy, the recession arises from the expansion of the universe: the speed of recession is directly proportional to the distance of an object, and hence red shift.

Reduced Mass

Reduced mass is the quantity, and it is denoted by $m_1m_2/(m_1 + m_2)$. In which, m_1 and m_2 are respectively the masses of two particles which apply equal and opposite forces on each other and are not subject to any external forces. The motion of either particle with respect to the other as origin is the same, with respect to a fixed origin, as that of a particle whose mass is the reduced mass and which is acted on by the same force.

Reflecting Telescope

Reflecting telescope is a kind of telescope in which the object is first imaged with the help of a curved mirror. This introduces no chromatic aberration and, provided a parabolic reflector is used, there is little spherical aberration either.

Reflection

Reflection is the process in which the radiation incident on a surface is returned by that surface to the medium of incidence, in spite of passing into the medium of the surface.

Refracting Telescope

Refracting telescope is a type of an optical telescope, just like the Kepler telescope and the Galilean telescope. Refracting telescope is consisting of two lens systems, the focal length of the objective system being greater as compared to that of the eyepiece system. Difficulty of grinding large lenses has resulted in the reflecting telescope being preferred for astronomical work.

Refractive Index

Refractive index is denoted by n (or sometimes μ). Refractive index is the ratio of the sine of the angle of incidence of electromagnetic radiation on a medium to the sine of its angle of refraction in the medium. The refractive index also equals to the ratio of the speed of the incident radiation to that of the refracted radiation.

If the medium of incidence is a vacuum, then in that case the index is said to be an absolute refractive index, and on the other hand, it is known as the relative refractive index to the medium of incidence. Refractive index varies with wavelength, resulting in Dispersion.

Relative Density

Relative density for solids and liquids is the ratio of the density of a substance at a stated temperature to the density of water at 4°C. In this way, the density expressed in gramme per cubic centimeter numerically equals to the relative density. Any substance which is having a relative density less than 1, it will tend to float on water. Usually, the relative density of a gas is stated with reference to the density of air at standard temperature and pressure.

Relative Velocity

Relative velocity is the velocity of one body A relative to another body B. At speeds small compared with that of light, A's velocity relative to B is found by compounding (that is, vectorially adding) A's velocity with B's reversed velocity. In simple words, A's velocity is compounded with a velocity which when applied to B would reduce B to rest.

In this way, for velocities of the objects who are moving in the same direction the relative velocity is obtained with the help of simple subtraction, and for velocities of the objects who are moving in opposite directions it is obtained by simple addition.

Relativity

Relativity is a theory that put forward to explanation for the failure of Newton's Laws of Motion when applied to very high-speed relative motion. The special theory of relativity refers to the nonaccelerated frames of reference. It considers that the physical laws are same in all such frames and that the speed of the light in a vacuum is constant all across the universe and it is independent of the speed of the observer.

Relaxation Time

Relaxation time is a measure of the time that has been taken to reach the state of equilibrium in a relaxation phenomenon. Relaxation time is the time required for an exponential function of time, 't' to reach '1/e' of its initial value. In this way, for the function exp (-at), in which, a is a constant, the relaxation time is '1/a.' This type of time development arises whenever the rate of change of a physical variable is proportional to the variable, as in Radioactive Decay.

Relay

Relay is a type of an electric device which utilizes a current variation in one circuit in order to control the current present in the other circuit. For instance, the current in the first circuit may energize an electromagnet which operates a switch in the second circuit. Other types of relay are consisting of devices like, the Thyratron. Electronic switching with the help of semiconductors is also employed.

Renormalization Group

Renormalization group is a method which is used in order to treating the systems which deals with all scales of length, from the atomic to the very large. It allows the Universality Class of a particular material and its Critical Exponent values to be anticipated. Never the less, the problem of critical phenomena in disordered systems remains the same.

Resistivity

Resistivity is denoted by symbol p. Resistivity is the quantity 'RAil,' in which R denotes the resistance of a conductor of the material under consideration, A denotes the uniform cross-sectional area, and l is the l ength. The SI unit of resistivity is the ohm meter.

Resistor

Resistor is a component which is used in an electric circuit to introduce the Resistance. Carbon resistors are extensively utilized in various electronic circuits, the value being indicated with the help of an arrangement and various rings are painted of different colors on each resistor. Generally, coils of wire, manganin, are also used as a resistor.

Figure 48. Resistor.

Source: Image by Flickr.

Rheostat

Rheostat is a variable resistor which is connected in series with a circuit to govern the current that passes through it. Generally, a rheostat is consisting of a coil which is of high resistance, a wire wound on a tube or ring, resistance variation being attained with the help of a sliding contact. The term is usually applied to physically large devices. A small rheostat is appropriate for electronic circuits, and is commonly known as a Potentiometer.

Ripple Tank

Ripple tank is a type of tank which is consisting of water, into which a vibrating strip or spheres just dip. Disturbances then spread out from the vibrators, the water depth being greater than half the ripple wavelength. Annotations can be made of straight waves from the strip or circular waves from a sphere, or the phenomena resulting when disturbances from the spheres meet.

Rocket

Rocket is a device which is propelled with the help of the thrust produced by the ejection of high velocity matter, that is, by Jet Propulsion. The rocket carries both its own fuel as well as the oxidant. In this way, the rocket is independent of the Earth's atmosphere for lift, thrust and oxygen. It is the only known vehicle in order to achieve flight in space.

Rotational Energy Levels

Rotational energy level is defined as, the allowed energies of rotation about axes through the center of mass of a molecule. With the help of the quantum theory, change from one rotational energy level to a lower one is escorted or assisted by the emission of a quantum of radiation.

Rotational Motion

Rotational motion is a type of motion, in which, when the angular acceleration is constant, is described by equations analogous to the four equations for linear motion with constant acceleration.

Rotation Spectrum

Rotation spectrum is a type of spectrum which is resulting from a change of a molecule between rotational energy levels in the same vibrational energy level. These sorts of spectra take place in the far infrared and microwave regions of the spectrum.

Rydberg Series

Rydberg series is a series of spectral lines in the absorption spectrum of a gas. In Rydberg series the separation between neighboring lines decreases with the decrement in the wavelength take place. Eventually they merge into a continuum at a point corresponding in order to complete the removal of the electron from the atom, that is, to ionization. Thus, the ionization potential may be decreases from the spectrum.

S

Sampling

Sampling is the selection and study of a small sample from a large collection. Sampling is thereby possible to attain both qualitative information as well as quantitative information about the collection as a whole.

Saturated Colors

Saturated colors are defined as the colors that are of visible monochromatic radiation. Mixing these sort of radiation with white light yields radiation of an unsaturated color but of the same hue as the monochromatic radiation. The proportion of white light is directly proportional to the unsaturation of the color.

Saturation Current

Saturation current is the maximum current, excluding breakdown current, which can be attained with the help of an electronic device, that is, the current value which cannot be increased by any increment in the voltage.

Saturation Force

Saturation force is the maximum value of the exchange force of attraction between protons in a nucleus. Protons in a nucleus can partake in Meson exchange with a finite number of other protons. Never the less, the electrostatic repulsive forces act between all the protons in the nucleus. Thus, with increasing size of nucleus, the saturation force increases more slowly as compared to the repulsive force.

Scalar Product

Scalar product is the multiplication of two vectors A and B say; it is written A.B and read as A dot B. It equals AB cos θ where A and Bare the magnitudes of the vectors A and B, and θ is the angle between the vectors.

For instance, work is the scalar produce of force and displacement. If in Cartesian Co-ordinates the vectors A and Bare (A_x, A_y, A_z) and (B_x, B_y, B_z), respectively, then $A.B = A/J_x + A_j J_y + A$.

Scaler

Scaler is an electronic device which generates an output pulse when a specified number of input pulses have been received. The main application is in conjunction with a Scintillation Counter in radioactivity measurement.

Scanning

Scanning is the methodical measurement or observation of some physical quantity at a series of positions on a surface or throughout a volume. Scanning may be continuous, as in the movement of a radar aerial scanning a surrounding volume, or intermittent, as in the line-by-line movement of an electron beam across the fluorescent screen of a television receiver.

Scattering

Scattering is a process in which the collision of a particle or radiation with another particle take place. This collision or discontinuity generates a change in direction or energy of participants in the collision. In inelastic scattering the total kinetic energy changes, on the other hand, in elastic scattering it is unchanged. In coherent scattering a definite phase relation exists between scattered and incident waves. In incoherent scattering no definite phase relation exists.

Schlieren Photography

Schlieren photography is a type of high-speed photography which is done with the help of sparking the illumination, of inhomogeneities in a transparent medium. The in-homogeneities are because of the density and refractive index variations in the medium. The variations show up in the form of streaks on the photographs taken by transmitted light.

Schmidt Camera

Schmidt Camera is a type of camera which uses a concave spherical mirror with a specially shaped lens in front of it to form the image. The lens is weakly converging near its center and weakly diverging near the periphery, in this way, it assists in correcting the system for spherical aberration. The resulting camera gives high light-gathering power which is combined with the sharpness of the image and hence, can be used in astronomical research or at high altitude photography.

Scintillation

Scintillation is defined as the light flash which is being generated by the fluorescence and which took place when a particle of suitable energy falls on an appropriate substance. That substance is known as a scintillator.

Scintillation Counter

Scintillation counter is a device which is used for counting the Scintillations. Sodium iodide activated with thallium is the generally used phosphor in scintillation counters. Commonly, each scintillation generates an output pulse in a suitably mounted Photomultiplier, and those pulses can be counted with the help of a Scaler. As the magnitude of the pulse relies on the energy of the particle that helps in producing the scintillation, use of a Pulse Height Analyzer allows a study of particle energies.

Scotopic Vision

Scotopic vision is a type of vision that has a very low luminance levels, when the main light receptors in the retina are cells known as rods. Scotopic vision is devoid of color sensation.

Scrambler

Scrambler is a type of circuit which is utilized in the transmission of signals, and renders the signals unintelligible on reception unless an appropriate circuit is used in the receiving equipment.

Screen Grid

Screen grid is a type of grid which is situated between the anode and control grid of a Thermionic valve. Screen grid held at constant potential to reduce the capacitance between the anode and control grid.

Sedimentation

Sedimentation is a routine method which is use to separate large particles from the solution or suspension with the help of a centrifuge. The particles end up in a hard pellet at the outer wall of the centrifuge and are said to have been sedimented.

Seismograph

Seismograph is a type of an instrument, which is used for the detection and recording of ground movement. Seismograph is consisting of a massive pendulum which is set in motion by the force produced at its suspension with the help of the ground movement, which could for example be generated by a nuclear explosion or an earthquake.

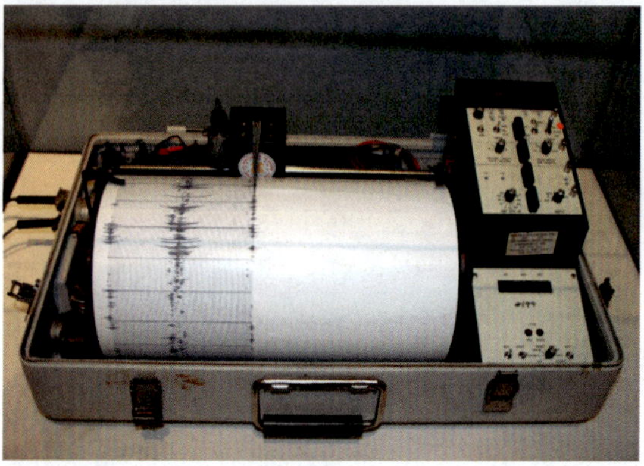

Figure 49. Seismograph.

Source: Image by Wikimedia Commons.

Seismology

Seismology is a study that deals with the Earth's structure from Seismograph recordings at various points on the Earth's surface of the same event: the time taken by waves from the focus, that is, the origin of the disturbance, in order to reach a detector relies on the nature of the medium by the means of which they travel.

Selenium Cell

A photocell in which a metal support has a thin selenium layer coated by a transparent gold coating. When light strikes the film, the photovoltaic effect creates a potential difference between the selenium and the gold.

Semiconductor Detector

A instrument that counts low-energy particles. It comprises of a reverse-biased P-N junction that swiftly sweeps away any electrons or holes created near the junction. The passage of a charged particle over the junction produces electron-hole pairs, which, if in sufficient quantities, result in a visible pulse along the circuit. Thc scnsitivity of a scmiconductor detector is approximately ten times that of a gaseous detector. Furthermore, due to its small size, it can be utilized in situations where other types of counters cannot be used. Because of its sensitivity to gamma rays and neutrons, it is ideal for counting in the presence of such background radiation.

Semiconductor Rectifier

A semiconductor rectifier is type of device, when an external potential difference opposes the junction's internal potential difference, the transport of electrons and holes across the junction is restarted, as shown. As a result, current flows, with the magnitude increasing as the voltage ıncreases.

Sensation Level

The symbol L represents the quantity 10 $\log(I/I_0)$, where I and I_0 represent the sound intensity under consideration and the minimum audible sound intensity, respectively. I_0 is commonly interpreted as 2.5 picowatts per square meter.

Servo Mechanism

A mechanism in which a small amount of power is utilized to control a much larger amount of power in a mechanical motion. feedback may be used in the system.

Shadow

When an opaque body blocks a portion of the light falling on a surface, a dark patch forms. The shadow is absolutely dark for a point source of light. For a long source, the shadow is composed of a completely dark region bordered by a region of partial shade. Except for diffraction phenomena at the edges, the nature of shadows obeys the laws of geometric optics; these phenomena are more noticeable the smaller the opaque body.

Shell Model

A theoretical model of the nucleus based on the notion that nucleons move in a central field, similar to how electrons in an atom move in the nucleus' field. Furthermore, quantum mechanics predicts that a nucleon moves in a shell similar to an atom's electron shell. The nucleus is regarded to be exceptionally stable if the amount of nucleons is just enough to fill an exact number of shells, i.e., it has magic numbers of protons and neutrons.

Shock Wave

When there is relative supersonic motion between a fluid and an obstacle, a relatively narrow zone of high pressure and temperature is generated in the fluid. A shock wave can also be generated in a fluid by a strong disturbance, such as a lightning strike or a bomb explosion.

Shunt

A low-value resistor connected in parallel with a piece of electrical equipment to limit the current flowing through it by a known factor. Shunts are frequently used to extend the range of ammeters.

Shunt-Wound

Electric motor winding in which the field coils are connected in parallel with the armature, resulting in low-speed change with load.

Sidereal Time

Time measured in terms of the stars. The sidereal year is the time between successive Sun-Moon conjunctions with a fixed star; the sidereal month is the time between two successive Moon-Moon conjunctions with a fixed star, and the sidereal day is the time between successive transits of a fixed star across the observer's meridian.

The sidereal year is 365.25636 days long; the sidereal month is 27.32166 days long, and the sidereal day is 23 hours and 56.06 minutes long. The gap between sidereal time and solar time is caused by the imposition of the Earth's orbital motion on the rotational motion.

Significant Figures

The number of reliable figures given for a numerical value of a physical quantity. It is important to note that the number of significant figures mentioned for a numerical value should never be greater than the smallest number of such numbers among all other variables used. Thus, if a mass of 1.52 kilogram is subjected to an acceleration of 2.141 meter per second per second, the force required should be 3.25 newton, not 3.25432 newton.

Silkscreen Process

A method of fabricating an integrated circuit on a ceramic substrate. The amounts of metals and other components in the inks utilized to govern the resistive qualities of the individual tracks.

Small-Angle Scattering

The scattering of waves at small angles by particles or cavities in a medium, where the wavelength is small in comparison to their dimensions. The most common kind is probably small-angle scattering of X rays by discontinuities, which is used in molecular studies.

Sodium Vapor Lamp

A gas discharge lamp in which the discharge begins in an inert gas such as neon at room temperature. This causes the sodium present to vaporize, resulting in the emission of the distinctive strong orange-yellow light. The use of neon gas explains the red color of the sodium vapor lamp when it is first turned on.

Solar Constant

At the Earth's average orbital distance from the Sun, the quantity of solar energy travelling typically through a unit area. The value would be 1390 watt per square meter if the Earth's atmosphere did not absorb it.

Solar Time

time is measured with relation to the Sun. The solar day is the period of time between successive transits of the Sun across the meridian of the observer. The mean solar day is calculated by taking the average of this time over a year: the mean solar day has 24 mean solar hours, each of which contains 60 mean solar seconds. The solar month is the amount of time it takes the Moon to complete one revolution of the Earth and return to the same longitude; it equals 27.32158 mean solar day.

Solar Units

Solar units are units that take the Sun's mass, diameter, density, and luminosity as a unit of measurement; these parameters for other stars are then stated in solar units.

Solar Wind

A plasma of protons and electrons that travels away from the Sun at speeds ranging from 300 to 800 kilometers per second. It is the driving force behind the formation of the Van Allen Belts and the Aurora. It also causes the Earth's magnetic field to extend deeper into space on the Earth's side away from the Sun. The solar wind is at its strongest when solar flares and sunspots appear.

Split Ring Commutator

An instrument that extracts unidirectional current from a generator. The top lead runs over the top half of the ring when the coil is rotating, while the bottom lead runs over the bottom half. The electromotive force reverses when the leads reach the gap, as do the leads on the split ring. As a result, the upper half's polarity with regard to the lower half remains constant, resulting in a unidirectional but changing electromotive force.

Staining

A method of dyeing sections of an item, usually on a microscope slide, so that, if the dye is selectively taken up, the intensity of the image varies much more strikingly from point to point than it would without the staining. The object's visibility is thereby improved, but there is one drawback: most staining procedures harm cells.

Standard Illuminants

Three sources that have been defined for usage in standard Colorimetry. They are of color temperature 2848 K, 4800 K, and 6600 K. When a sample that is not self-luminous is lit at 45° to the normal by one of these sources and viewed normally, the color coordinates are determined.

Standards

Quantities described in terms of physical conditions that are thought to be constant. Some previously adopted standards suffered from the drawbacks of change. Corrosion, crystallization, and other processes, for example, influenced the length of the metal block once designated as the standard of length.

Standard Temperature and Pressure

0°C, or 273.15 K, and 101 325 pascal, respectively; the latter figure roughly approaches the previous standard pressure of 760 millimeters of mercury.

For the properties of gases, standard temperature and pressure are standard conditions.

Star

Any of a vast number of massive astronomical objects that emit electromagnetic radiation and are organized into galaxies. Even within galaxies, the stars, of which the Sun is one, are separated by huge distances and hence appear only as points of light from Earth.

State of Matter

Any of the three physical states in which matter can exist, namely solid, liquid, and gas. Changes in state occur at certain transition temperatures that are unique to each substance. Plasmas, liquid crystals, glasses, colloids, and super fluids have all been described as additional states of matter at various times.

Stationary Orbit

A circular synchronous orbit in the same plane as a celestial body's equator, generally the Earth. A satellite in this orbit seems to bc stationary in rclation to a location on Earth. Communication satellites in stationary orbit reflect radio waves, and by combining many of them, global coverage is achieved.

Statistical Mechanics

The statistical prediction of the properties of matter in bulk based on knowledge of the constituent atoms or molecules and the forces that exist between them. The terminology is sometimes used to refer to the treatment of systems that are in equilibrium.

Statistical Weight

Symbol g. A number equal to the number of times a specific value, or range of values, of a quantity occurs in a statistical analysis of the quantity.

Steady State Theory

The hypothesis that the universe has always existed, with its expansion compensated for by the continual creation of matter at a rate of roughly IQ–43 kilogram per cubic meter per second, hence maintaining the universe's average density constant. Many observations, however, suggest to favor the alternative Big Bang Theory.

Steam

At a temperature of 100°C or above, water is in a gaseous state. Steam is technically an invisible gas, but the phrase is also given to steam that carries minute liquid water droplets and hence appears white.

Steam Calorimeter

A calorimeter that calculates the quantity of heat delivered based on the mass of steam condensed on the body under test.

Steam Engine

A machine that takes heat from a steam boiler, uses it to do external work, and then rejects the unused heat to a condenser.

Figure 50. Steam Engine.

Source: Image by Wikimedia Commons.

Stefan's Law

The law $M = \sigma T^4$ where M is the energy radiated per unit area per second from a black body at thermodynamic temperature T; σ is Stefan's constant and is provided by $2\pi^5 k^4/(15h^3c^2)$, where k is the Boltzmann Constant, c is the speed of light in vacuum, and h is the Planck Constant. 5.6697×10^{-8} watt per square meter per kelvin to the fourth is the value of σ.

Stellar Evolution

The gradual series of changes that occur in a star as it ages. Stars of similar mass to the Sun evolve into a Red Giant and then a White Dwarf. Very huge massive stars eventually become unstable, resulting in a supernova explosion and the star's core collapsing to create a Neutron Star or Black Hole.

Stellar Spectrum

The spectrum of radiation emitted by a star. It reveals information about the temperature and composition of the star. The shorter the wavelength.

Synchrotron Radiation

Electromagnetic radiation created when electrons move quickly in a magnetic field. Electrons move in a circular pattern. The radiation is released in the direction of the electron's velocity and is plane polarized perpendicular to the magnetic field direction.

T

Tachyon

A hypothetical elementary particle with a speed faster than the speed of light, such that, according to relativity theory, either its rest mass or its energy is fictitious. In a vacuum, such a particle should emit Cerenkov Radiation. Tachyons are still undetected.

Tangent Galvanometer

A galvanometer in which the current to be measured is routed through an n-turn vertical wire coil, the plane of which is parallel to the horizontal component H of the Earth's magnetic field. A tiny magnet is pivoted in the center of the coil, allowing it to move horizontally. As a result of the magnetic activity between the coil's and Earth's fields, it is deflected through an angle of 9.

Tau Lepton

In the third generation, a negatively charged lepton. It is a heavier version of the electron and the muon.

Telecommunications

The transmission of information using any type of electromagnetic system, such as a phone, radio, television, or telegraph.

Telegraph

A device that sends and receives messages by delivering electric pulses along cables.

Telemeter

Any device that sends measured data from its source to a remote observer. Thus, measurements of conditions within a nuclear reactor core are communicated to the controller via cables.

Telephone Lens

A photographic compound lens that consists of a converging lens system followed by a diverging lens system, resulting in an actual image with larger magnification than the standard camera lens that it substitutes when needed.

Telescope

An optical instrument used to create a magnified image of a distant object. A refracting telescope employs a combination of lenses, whereas a reflecting telescope has at least one reflecting surface and may additionally employ lenses. A terrestrial telescope is one that produces an erect final image, even if it is ideal for astronomical operations.

Figure 51. Telescope.

Source: Image by Pixabay.

A telescope that produces an inverted final image is referred to as an astronomical telescope. A telescope is said to be in normal adjustment when both the object and the picture generated by it are at infinity, i.e., parallel light strikes and emerges from the instrument.

Television

The electrical transmission of visual sequences and images via cable or radio waves. The transmitted scene is optically focused on a photoelectric screen made up of a mosaic of light-sensitive particles in a camera tube.

Temperature

The letter T is used as a symbol for Temperature. A unit of measurement for a body's heat. The direction of heat flow is determined by the temperature, which is always from higher to lower. A body's temperature is a measurement of the average kinetic energy of its atoms or molecules.

Terminal Velocity

When the resultant force on the body is zero, the velocity of a body relative to the fluid in which it is travelling. The terminal velocity of a sphere of radius r and density p falling under gravity in a fluid of density σ is given by $2(\rho - \sigma)r^2g/(9n)$.

Terrestrial Magnetism

The magnetism of the Earth. The Earth acts as if it were surrounded by a huge bar magnet at a little angle to the Earth's axis of rotation. The magnetic North and South Poles, which do not correlate with the geographic poles, have the strongest magnetic fields. Furthermore, the magnetic pole positions fluctuate slowly over time. The horizontal and vertical components of the magnetic field, as well as the angles of dip and declination, characterize the magnetic field at places on the Earth's surface.

Tetrode

A thermionic valve with four electrodes. It is thus a triode with an additional grid, known as a screen grid, between the anode and control grids, lowering capacitance between them and thus boosting the valve's performance as a high frequency amplifier.

Thermal Agitation

The random movement of a substance's molecules. The energy of motion is the heat component of the substance, according to the kinetic theory of matter.

Thermal Conductivity

Symbol A. The amount of heat flowing per second across a unit area of a substance given a unit temperature gradient, with the plane of the area perpendicular to the temperature gradient.

Thermal Diffusion

When a mixture of gases is subjected to a temperature gradient, the heavier atoms or molecules concentrate in the colder part, while the lighter ones concentrate in the hotter region. Some isotopes are separated via this phenomena.

Thermal Diffusivity

It is the thermal conductivity divided by density and specific heat capacity at constant pressure.

Thermionic Emission

By heating a solid the release of the electrons obtained. To overcome the solid work, function the effect occurs when significant numbers of electrons have sufficient kinetic energy. With the temperature the effect increases.

Thermionic Valve

Using a thermionic cathode, a multielectrode evacuated electron tube using as electron source. To the anode current only flows when a positive potential is applied. applying a signal to one of the extra electrodes causes modulation of the anode current and hence amplification of the signal, with three or more electrodes. Values have now been superseded by their solid-state equivalents for most applications with the three or more electro lodes and semiconductor diodes, field effect transistors etc. Therefore, some special-purpose valves are still important.

Thermistor

With the temperature a piece of semiconducting material whose resistance falls rapidly. It has been seen that temperature changes as small as 0.005°C can be measured by using a thermistor as one arm of a suitable electrical bridge. On heating thermistors are also used to compensate for the increased resistance of ordinary resistors. Other uses are in vacuum gauges, time delay switches and voltage regulators.

Thermocouple

For measuring a thermoelectric device is being used. At both ends two dissimilar metal wires or semiconducting rods are joined. In the production of an electromotive force a temperature difference between the junctions results in the production and the phenomenon is known as the See beck effect.

By a millivoltmeter or potentiometer incorporated in the circuit the electromotive force may be measured. One junction is maintained at a constant known temperature and the other at the temperature to be measured for temperature measurement which can be obtained from a calibration curve of voltage against junction temperature difference.

Thermogram

From a body an image is being produced by the thermal radiation. To a black body it is a valuable diagnostic aid since human skin, independent of its color, is a good approximation and by circulatory and cellular abnormalities marked changes in the radiation temperature are produced and are thus shown up on the thermogram. In breast cancer screening the method has been particularly successful.

Thermoluminescence

When certain solid materials, containing electrons trapped at defects, are heated when luminescence produce. The effect of heating is to free the electrons, which then fall to lower energy states with the emission of photons. The phenomenon is used for example in pottery dating.

Thermometer

For measuring the temperature, it is an instrument. There are many different types, each depending on the easy measurement of some physical property that varies with temperature. Examples, mercury in glass thermometer, alcohol meter, constant pressure gas thermosteric, constant volume gas thermometer and also the resistance thermometer.

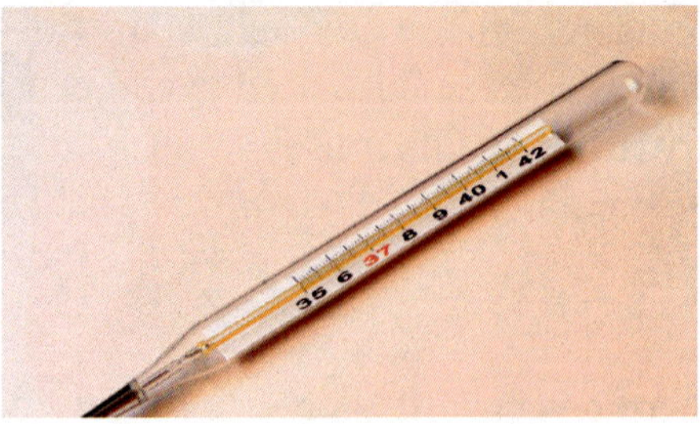

Figure 52. Thermometer.
Source: Image by Max Pixel.

Thermophone

Between two terminal blocks a thin metal strip of small heat capacity is being mounted. Through the strip, periodic temperature and on passing current through the strip variations are produced in it and these cause the expansions and contractions of the surrounding air, that is sound waves. The sound output requires amplification before use.

Thermopile

For measuring the radiant heat this is an instrument. In series it consists of a thermocouple collection connected and the alternate junctions, blackened to increase absorption, are exposed to the radiation, the other junctions being shielded. To the radiation intensity a measurable electromotive force, proportional and thus produced.

Thermostat

For maintaining a constant temperature, a device is usually in an enclosure. To a switching device it consists of a temperature sensing instrument connected. A simple electric switch the sensing device is frequently a bimetallic strip which triggers a simple. A device for maintaining a constant temperature, usually in an enclosure. It to a switching device. To describe the constant temperature enclosure this term is sometimes used.

Thomson Effect

When a temperature gradient is maintained along a conductor through which a current is passed, heat is evolved at· a rate roughly proportional to the product of the current and the temperature gradient is the phenomenon. If either the gradient or the current is reversed, heat is absorbed rather than evolved. The phenomenon is superimposed on the normal heating effect of a current.

Thomson Scattering

By free or loosely bound electrons the scattering of electromagnetic radiation. Provided that the incident photon energy is much less than the rest energy of the electron, the total intensity of the scattered radiation is given by 8nr2//3 where r is the classical electron radius and I the intensity per unit area of the incident radiation.

Thyratron

In the particle counters and relays a type of thermionic gas-filled TRIODE used. In the grid potential current flow is being initiated and it then becomes independent of grid potential and can only be stopped by a suitable reduction of anode potential. The thyratron is now being superseded by solid state devices.

Tides

In the Earth's oceans the regular rise and fall of the level of water due to the gravitational forces of the Sun and Moon on the Earth and the effect of the moon is approximately twice that of the sun. According to the Moon's position relative to the Earth the vector sum of the forces acting varies according to, high water has different heights at different times. Tidal power stations using specially designed tidal water operated turbines to drive generators and so produce electricity have been built at several sites.

Time

With some agreed process the duration between two events as determined by comparison. It has been said that usually a periodic process is chosen: examples are the swing of a pendulum, the vibration of electromagnetic radiation or various astronomical processes. The calendar year has an average length equal to that of the solar YEAR. So that the calendar year may contain an exact number of days, each year has 365 days with a leap year containing 366 days every fourth year: only century years yielding an exact integer when divided by 400 are leap years, thus 2000 will be a leap year but 1900 was not.

Torsional Vibration

In one direction a vibration associated with a twist occurring first and then the other. For instance, by a vertical wire if a horizontal disc is suspended from its center which is firmly clamped at the other end, and the disc is Rota ted in its own plane and then released, it performs torsional vibrations due to the twisting of the wire first in one direction and then the other. By timing the oscillations, the rigidity modulus of the wire can be found.

Total Reflecting Prism

A right-angled crown glass isosceles prism. When light is normally incident on it, as shown and the light strikes the hypotenuse face at an angle of 45°; this is greater than the critical angle, whose value for crown glass is 41 o 30.' The light is thus totally internally reflected and emerges as shown. The advantages of such a prism over a mirror are that multiple images are avoided as is also the 10% intensity loss occurring at a silvered surface.

Transformer

By using electromagnetic induction, a device for changing alternating-current power from one voltage or current level to another without frequency change. The transformer consists essentially of two coils of wire round a ferromagnetic core. Power fed into the primary winding produces a changing magnetic field which induces power in the secondary winding.

Transistor

By the potential applied to a third electrode a device composed of semiconductor material in which the flow of current between two electrodes is controlled. In addition to rectification, it is thus capable of amplification. It is the basic unit of radio, television and computer circuits and has almost completely replaced the thermionic valve.

Tyndall Effect

By the small particles in its paths the scattering of light, therefore causing the path of a light beam in, for example, a dusty atmosphere to become visible.

U

Ultracentrifuge

At a very high angular speed a centrifuge designed to work, so that the centrifugal force produced is sufficiently large to cause sedimentation of colloids. On the particle size the rate of sedimentation depends and this can be investigated. Relative molecular masses of proteins and other macromolecules can also be measured. In quantitative instruments, the formation of sediment is photographically recorded.

Ultraviolet Radiation

In the wavelength range of about 10 nanometer to 380 nanometer there is the electromagnetic radiation that is which is lying between X ray and visible light wavelengths. The ultraviolet region is arbitrarily divided into the near ultraviolet, wavelength range 200 nanometer to 380 nanometer, and the far ultraviolet, also known as the vacuum ultraviolet.

Ultraviolet Spectrum

In the ultraviolet region an emission or the absorption spectrum. On the energy levels and molecular structure is the spectra yield information. In the ultraviolet spectroscopy optical components of glass cannot be used in because of their strong absorption. In the near ultraviolet quartz components are suitable; at shorter wavelengths vacuum mounted reflection gratings are employed.

Unit

As being the unit amount of that quantity there is an amount of a physical quantity which is arbitrarily defined. For example, a particular piece of platinum has been assigned a mass of one kilogram and is the unit of mass in SI units. Any measurement is effectively a comparison of what is being measured with the appropriate standard, which may be a Primary Standard or a Secondary Standard depending on the accuracy required and on convenience.

Unitary Symmetry

With the SU(3), a theory which is concerned with and predicts that, with respect to strong interaction, elementary particles can be grouped into multiples containing 1, 8, 10 or 27 particles and that the particles in each multiples can be regarded as different states of the same particle. By plotting hypercharge against Isospin Quantum Number, the theory is usually illustrated graphically. The theory predicted the existence of the Omega Minus Particle prior to its discovery, but has not been completely successful.

Universe

In this the whole of space which matters and the mass of matter is estimated at 1041 kilogram and the age in the range 5 x 1 ()9 year to 50 x 1 ()9 year. For the speculation the origin is a matter although the Big Bang theory is now widely accepted.

Uranium

It has been symbolized by U and a metallic radioactive element of which 99% is uranium–238, 0.7% uranium–235 and the remainder uranium–234. Uranium–238, the most stable isotope, is used in breeder reactors as a source of the plutonium–239 isotope. Uranium enriched with uranium–235 is used as a Nuclear Reactor fuel.

Figure 53. Uranium.

Source: Image by Wikimedia Commons.

Uranium Dating

For the rocks it is considered as the type of radioactive dating. It has been thought that when rocks were formed only uranium238 was present and that the lead–206 now found is the product of the radioactive decay of the uranium during the intervening period. Therefore, by determining the amounts of uranium–238 and lead–206 now present in a rock sample, the age of the sample can be calculated.

Uranus

It is the planet which is seventh closest planet to the sun. With a dense atmosphere of hydrogen and methane it is giant planet and was the first to be discovered telescopically. Its orbital period around the Sun is 84 year and its axial rotation period is about 10.75 hour. The temperature is probably about –200°C. It has several satellites and a dark ring system. It has been investigated by a planetary probe.

Useful Angular Magnification

At the limit of resolution that the angular magnification i.e., magnifying power, of an optical instrument for which both eye and instrument are working. Larger values of the angular magnification will make objects appear larger but will not increase their definition. For example, the limit of resolution of a telescope for light of wavelength 6×10^{-7} meter is $1.22 \times 6 \times 10^{-7} /D$ radian.

V

Vacuum

It is considered as the space in which there are relatively few atoms or molecules. In practice a perfect vacuum would contain none but is unattainable because of the finite vapor pressure of the enclosing walls. For pressures down to 10^{-2} pascal, the vacuum is said to be soft or low; from 10^{-2} pascal to 10^{-7} pascal, the vacuum is said to be hard; for pressures less than 10^{-7} pascal, it is said to be an ultra-high vacuum.

Vacuum Ultraviolet

About 200 nanometer the part of the ultraviolet region of the spectrum, of wavelengths are below in which the radiation is absorbed by air and so experiments using such radiation are performed in an evacuated chamber.

Valve Voltmeter

It is an amplifier which incorporate the value components. The output current, which is proportional to the input voltage, is read on a meter. The instrument uses very little input current because of its high input impedance. It is suitable for either alternating current or direct current input. The valve voltmeter is being superseded by a voltmeter with digital display.

Vapor Density

Under the specified conditions the density of a vapor, or sometimes the density relative to hydrogen under the same conditions. Since the latter is twice the vapor the vapor density measurement provides a method of determining relative molecular mass.

Vector Addition

By two sides of a parallelogram drawn from one corner in the case if vectors A and B are completely represented then the vector A + B is completely represented by the diagonal of that parallelogram drawn from the same corner.

Vector Field

At any point is completely a region of space for which a particular property – magnetic, electrostatic etc. which is completely described by a vector. By the lines such a field can be mapped whose density at any point (i.e., the number of lines per unit area crossing an infinitesimally small area perpendicular to the lines) is proportional to the magnitude of the vector at that point and whose direction is that of the vector.

Velocity

In a particular direction the rate of increase of distance traversed by a body. Velocity is thus a vector quantity. The velocity of a particle travelling in a curved path has the direction of the tangent to the path at the point considered.

Velocity Modulation

From a continuous stream of them the production of pulses of the charged particles. By passing the stream into a sharply defined region such as a cavity resonator and there subjecting it to a rapidly fluctuating alternating electric field it is accomplished. According to the nature the field of the field half cycle when they enter it will either accelerate or retard the particles. For the particles immediately following, the field is reversed and so the effect on them is the opposite. The particles are thus bunched and proceed in pulses. Velocity modulation is used for the amplification and generation of microwave frequencies in electron tubes such as the klystron and the travelling wave tube.

Velocity Potential

At aby point a scalar function of the position of a moving particle such that its gradient gives the particle velocity. It has applications for example to fluid flow and to the movement of electric charge.

Velocity Selector

It is a device in which an electric and a magnetic field act at the right angles to each other. By varying the strengths of the fields, it can be arranged that all charged particles, except those of a chosen velocity, can be deflected from the initial path of a stream of mixed velocity particles.

Vibration

It has been said that originally it is an elasticity of the body and the word is now used synonymously with oscillation. In longitudinal vibration the particles move to and from along the line of travel of the wave. In transverse vibration the particles move to and from in a direction perpendicular to the line of travel of the wave. Reason ant vibrations occur when a periodic force of the same frequency as the natural frequency of a system is applied to the system. The amplitude of the vibrations may become very large.

Vibrational Energy

It is considered as the sum of the kinetic and potential energies of an oscillating body. For undamped simple harmonic motion, the sum is constant and equal to $mw^2a^2/2$; m is the mass of the body, a its amplitude of swing (assumed small) and $2\pi/w$ its period. At maximum displacement all the energy is potential; for zero displacement it is all kinetic.

Vibrational Energy Level

About its equilibrium position the value of the vibrational energy of an atom. According to quantum theory only certain vibrational energy levels are allowed and changes in vibrational energy only occur from one allowed level to another.

Vibration-Rotation Spectrum

From a change of a molecule from a rotational energy level in one vibrational energy level to a rotational energy level in a different vibrational energy level which a spectrum resulting. It has been said that such spectra occur in the near infrared region.

Vidicon

To a photo conducting mosaic coated on an insulating screen it is a type of television camera in which the image is projected on to, which is scanned by a low-velocity electron beam; a positive potential is applied to the mosaic. On the amount of illumination, it receives the amount of charge on each mosaic element depends. This charge is released by the electron beam, thus producing a current which depends on the amount of charge and therefore on the amount of incident illumination.

Virtual Particle

For a very short time a particle is also existence and whose continued existence would violate the conservation law for mass and energy: very short-term violation is allowable because of the heisenberg uncertainty principle. An example is the emission and absorption of virtual photons during an electromagnetic interaction between two charged particles.

Viscometer

For measuring the fluid viscosity, it is an instrument or apparatus. The main types are based on (a) the flow of fluids through capillary tubes; (b) the time of fall of a sphere through a liquid; (c) the torque required to keep two concentric cylinders rotating when the space between them is occupied by the fluid under test; (d) the rate of damping of a vibrating body by the fluid.

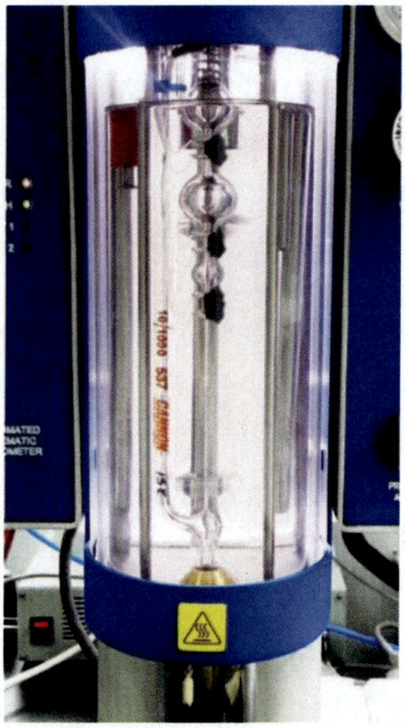

Figure 54. Viscometer.

Source: Image by Wikimedia Commons.

Visible Spectrum

It is considered as the electromagnetic radiation in the wavelength range 380 nanometer to 780 nanometer. On the horseshoe part of the chromaticity chart boundary, it is represented by the points. This boundary radiation broadly appears to divide into seven colors: red, orange, yellow, green, blue, indigo and violet (mnemonic: Richard of York Gained Battles in Vain).

The wavelength ranges in nanometer for these colors are respectively 780–620, 620–585, 585–575, 575–500, 500–445, 445–425, 425–380. In reality there is a continuous gradation of color through the spectrum. The line joining the extremes of the horseshoe represents purples, i.e., not spectral colors.

Voigt Effect

By passing light through a vapor subjected to a magnetic field the double refraction produced, the field being in a direction perpendicular to the direction of incidence of the light.

Volt

It is symbolized by the letter V. The SI unit of electrostatic potential, of potential difference and of electromotive force. It is defined as the potential difference between two points on a conductor carrying a steady current of one ampere, when the power dissipated between the points is one watt.

Voltage Transformer

To the main circuit and to the instrument respectively an instrument transformer whose primary and secondary windings are connected. From the high voltage circuits such transformers are used to extend the range of alternating current instruments and to isolate instruments.

Voltmeter

For measuring the voltage, it is the device. It is one of the common forms is the permanent magnet moving coil instrument, which is suitable for direct voltages only. The Digital Display voltmeter, Cathode Ray Oscilloscope and Electrostatic Voltmeter are also used. All voltmeters are required to take little current from the circuit and so a series resistance is used with the moving coil instrument; cathode ray oscilloscopes take no current from the circuit.

Volume Compressor

For a large-amplitude input signal it is a device which automatically decreases the circuit amplification and increases it for a small one. It is a device which is known as the volume expander. Use of a compressor at the transmitter and an expander at the receiver improves the signal-to-noise ratio of a transmission system.

Volume Density of Charge

The electric charge per unit volume of a medium or body; it is measured in coulomb per cubic meter.

W

Wall Effect

On the behavior of the contents, it is an appreciable effect of the inside wall of a container or reaction vessel, for example a contribution to the current in an ionization chamber by electrons liberated from its walls.

Water

About three-quarters of the Earth's surface a colorless, odorless, tasteless liquid which covers and occurs in all living matter and many minerals. For the several physical quantities it is being used as the standard. Water has some unusual properties: thus, ICE is less dense than the liquid at 0°C, the respective values being 916.8 kilogram per cubic meter and 999.8 kilogram per cubic meter. Water is an excellent solvent for many compounds.

Wave Function

In the Schrodinger equation the function is appearing and also representing the amplitude of a wave associated with a particle. Its physical significance is that the square of the value of 'II at any point is proportional to the probability of finding the particle at that point.

Wave Mechanics

It is considered as one of the forms of quantum mechanics which is arising from the theory that a particle can be regarded as a De Broglie Wave, so that any system of particles, for example electrons orbiting atomic nuclei, can be described by Schrodinger's Equation.

The basic problems in wave mechanics are thus those of formulating and solving Schrodinger's equation for the system under consideration. In general, it is found that a particle can have only certain allowed energies and associated wave functions; the latter give the probabilities of finding the particle at different points in space.

Wave Meter

For measuring the frequency of the radio waves, it is considered as an instrument. It comprises a circuit which can be tuned to different frequencies, usually by means of a variable capacitor calibrated in frequency, and a device which indicates correct tuning.

Weak Interaction

About 10–12 times it is an interaction between elementary particles of magnitude that of strong interaction. It has been said that weak interaction force is considered to have a very short range. If both strong and weak interaction are permissible, only the strong will occur. Beta decay and the decay of some mesons and hyperons are examples of weak interaction. Weak interaction proceeds slowly on the nuclear time scale.

Weber

At a uniform rate the flux is being reduced to zero and the SI unit of magnetic flux, defined as that flux which, when linking a circuit of one turn, produces an electromotive force of I volt when, over a period of I second.

Weight

Due to the gravitational attraction of the Earth the force is being exerted on a body. Due to the gravity its direction is from the body to the Earth's center and its magnitude is mg where m is the mass of the body and is the magnitude of the acceleration.

Weight Thermometer

With a narrow tube attached it is a small glass or silica bulb which is used to determine the coefficient of expansion of a liquid. By a known amount the thermometer is filled with the liquid under investigation; the liquid temperature is then raised and also the mass m of expelled liquid and mass m' of remaining liquid determined. The apparent coefficient of cubical expansion of the liquid is then given by m/(m't).

White Light

It is known as the light which produces a color sensation of white. By the radiation such the sensation is being produced by having an energy spectrum, but can also result from radiation of many other spectral distributions, for example a mixture of red, green and blue monochromatic radiation. The definition of white light in terms of any particular spectral distribution is therefore to be deprecated.

Wimshurst Machine

An electrostatic generator consisting of two parallel insulating plates rotating in opposite directions about a common axis. Attached to each plate are radial conducting strips, for one plate; these strips pass in tum past stationary brushes BB (two for each plate). Initially one radial strip is charged; then rotation is started and charges of opposite sign are induced on strips of the other plate as they pass. Charges on both plates are collected by pointed combs at PP where a large potential difference can be built up.

Figure 55. Wimshurst machine.

Source: Image by Wikimedia Commons.

Wind Power

It is known as the utilization of the Wind Energy. For converting wind energy to electric power there are various types of wind mill which are being tried. Like the solar heating and wave power is a renewable source.

Wind Tunnel

It is a hollow tube through which passes a uniform flow of air. A scale model of the object under test, such as a car, aircraft or missile, is suspended in the tunnel. Deductions about the actual behavior of the object can be made from that of the model.

Wobbulator

Through a definite range of the values, it is a signal generator whose output frequency varies periodically. It is used to investigate the frequency response of circuits.

Wollaston Prism

By cementing together two triangular prisms of quartz or calcite a prism is being made. The optical axis of prism P1 is parallel to AB; that of prism P2 is perpendicular to the plane of the paper. The prism separates the ordinary and extraordinary components.

Wollaston Wire

A thin platinum wires whose diameter has a lower limit of one micrometer. It is made by drawing out a silver-encased platinum wire of larger diameter and then dissolving away the silver with acid.

Work

The product of the magnitude of a force and the distance moved by the point of application of the force in the direction of the force. Work done on a system increases the energy of the system. Work done by a system decreases the system's energy.

Work Function

The minimum energy required to remove an electron from a solid at the temperature of absolute zero.

Wrench

A couple together with a force which acts along the axis of the couple. The moment of the couple divided by the magnitude of the force is the pitch of the wrench and has the dimension of length. Any system of forces can, in general, be reduced to a wrench.

X

Xerography

It is considered as an electronic method of document copying. Ultraviolet radiation falls on an electrostatically charged plate, usually selenium coated after striking the document to be copied. With the amount of ultraviolet energy falling on it the loss of charge of any plate area increases. It has been said that a dark powder, consisting of thermosetting resin and graphite mixture and bearing a charge opposite to that of the plate, is then applied to the plate. Since like charges attract, the greater the charge on any plate area, the greater the amount of powder adhering there; hence a positive image of the document results.

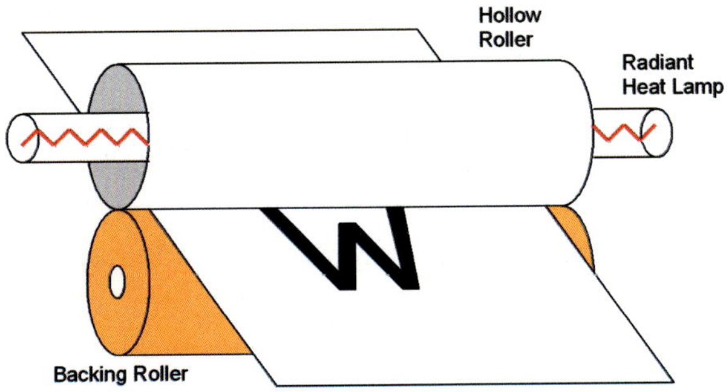

Figure 56. Xerography.

Source: Image by Wikimedia.

X Ray Absorption

It is considered as the process which usually involve either the Photoelectric Effect or Compton Effect. For the former, an X ray photon is completely annihilated, part of its energy being used to free an electron from an atom or molecule and the remainder giving the freed electron kinetic energy. In the Compton effect, the X ray photon encounters an effectively free electron and loses part of its energy to it. The photoelectric effect is predominant for low energy photons and rapidly increases with atomic number.

X Ray Analysis

By studying the X ray diffraction patterns, they produce the study of the structure of crystalline substances is being done. It has been said that either a collection of small crystals or a single large crystal may be used. In some equipment, the crystal may be rotated so that diffraction at successive sets of crystalline planes can be studied. Diffraction may be investigated by either transmission or reflection.

X Ray Astronomy

In the space it is the study of sources of X rays. Since X rays are absorbed by the Earth's atmosphere it is necessary to mount the instrumentation in space probes. It has been seen that the observations show a continuous X ray background spectrum with numbers of superimposed discrete X ray lines.

X Ray Diffraction

By the crystals the diffraction of X rays is usually the diffraction produced. It occurs because the wavelengths of X rays and the interatomic distances in crystals are of the same order of magnitude. In addition to it, the periodic crystal lattice acts like a diffraction grating.

X Ray Spectrometer

For measuring the variation of energy distribution of X rays with wavelength it is an instrument. It is considered as the crystal or fine diffraction grating disperses the X rays and the energies of the resulting beams are measured by ionization chamber, counter or photomultiplier or by photography.

X Ray Spectrum

It has been seen that it is a combination of a continuous and a superimposed line spectrum. In groups characteristics the sharp lines occur and the transfer of outer electrons to each of the various inner shells, i.e., K L, M etc.. By the electron the transfer may be investigated and also the X ray or gamma ray bombardment. The continuum arises from the retardation of charged particles, such as electrons stopped by the target of an x-ray tube; there is therefore a sharp cut off at the shortest wavelength present, corresponding to the energy of the bombarding particles.

X Ray Therapy

With the X rays the bombardment of tumors. For tumors on or near the skin an X ray tube operating at not more than 250 kilovolt is used; for deep-seated tumors the required operating voltage is around 4 megavolt. Since all cells along the path of the beam are irradiated, the beam direction relative to the body is altered for successive treatments. but always in such a way that the tumor lies in the beam path; hence no cells except the tumor ones receive large dosages. Higher energy X rays penetrate soft tissue better and are less absorbed in bone than lower energy X rays.

X-Y Recorder

With the two inputs it is a type of pen recorder and one displaces the pen in the x direction and the other in the direction, permitting a graph to be plotted of one varying electrical signal against another.

Y

Yellow Spot

In the retina of the eye, it is a small depression where the sensitivity of the eye is greatest. It has been seen that the yellow spot contains only close-packed cones and so has the best color discrimination and visual acuity.

Yield Point

For a substance corresponding to a marked change in the internal structure of the substance due to slipping of crystal planes the point on the stress against strain graph. It is characterized by a large change in the slope of the graph from its constant value at low stresses.

Ylem

About 1016 kilogram per cubic meter a hypothetical substance of density, consisting chiefly of neutrons. It may have been the ancestor of all nuclei.

Yoke

A fixed piece of unwound ferromagnetic material which completes a magnetic circuit

Young's Fringes

By monochromatic light of wavelength, A and so slits A and B fringes obtained using the apparatus which act as coherent equal-intensity sources. In the formation of interference fringes as indicated interference between the beams from these sources results.

Young's Modulus

The ratio of the tensile stress on a material to the resulting tensile strain for stresses smaller than the elastic limit.

Yukawa Potential

In the atomic nucleus it is a potential which is used to explain the short-range strong interaction occurring between nucleons. The potential energy is assumed proportional to $(\exp(-\mu r))/r$ where r is the nucleon separation and μ is a constant.

Z

Zeeman Effect

In a strong magnetic field, the splitting of spectral lines into components of slightly different wavelength when the light source is placed. On the direction the observed state of polarization of the components depends, relative to the magnetic field, from which the source is viewed. Treating the electron as a classical oscillator leads to the prediction that there should be two lines, circularly polarized in opposite directions, for each original line when viewing the source in the direction of the magnetic field.

Zener Breakdown

In a reverse biased P-N JUNCTION it is a type of breakdown which is found, particularly if the doping level is very high. From the valence band to the conduction band such a junction has a high built-in potential across it and the depletion layer is narrow, so that electrons may be excited directly. Breakdown is accompanied by a very sharp current increase.

Zener Diode

When the reverse voltage reaches a certain value a semiconductor diode for which a sudden increase in reverse current occurs, usually less than 6 volt Such diodes are used for regulating the voltage in a circuit.

Zero-Point Energy

At the temperature of absolute zero the energy of vibration of an atom is. Its value is hv/2 where h is the plank constant and v the frequency of vibration. In theoretical calculations this is taken to occur at infinity. In contrast for most practical work the surface of the Earth is considered to be at zero potential energy.

Zone of Silence

From a given source do not penetrate a localized region where sound or radio waves. Beyond the zone the waves are either reflected or refracted away from the region and may be received at some distance from the source.

Zone Plate

With the annular concentric zones which are alternately transparent and opaque and have radii such that the zone areas are approximately equal. For some particular observation point, the distance of the periphery of the zone from the point increases by half a wavelength from zone to zone; the contributions from each transparent zone are hence in phase at the point and so the intensity of illumination there is much greater than in the absence of the zone plate.

Zone Refining

For purifying materials, it is considered as the technique. Its main use is in semiconductor manufacture. A sample in bar form is slowly moved past a heater so that a molten zone passes along the length of the bar. Impurities concentrate in the melt and so are transferred to one end of the bar. By using several passes, impurity levels as low as one part in 1010 can be achieved.

Zoom Lens

In cinematic and television cameras an adjustable compound lens mainly used, whose magnification is continuously variable without loss of image sharpness.

Z Particle

In the weak interaction a massive electrically neutral particle which takes part.

BIBLIOGRAPHY

1. *A Dictionary of Physical Sciences*. (1976). [online] Available at: https://link.springer.com/book/10.1007%2F978–1–34 (accessed 4 August 2021).
2. *A Dictionary of Physics*. (2009). [online] Available at: https://www.oxfordreference.com/view/10.1093/acref/9780199233991.001.0001/acref–9780199233991 (accessed 4 August 2021).
3. *A Dictionary of Physics*. (2019). [online] Available at: https://www.oxfordreference.com/view/10.1093/acref/9780198821472.001.0001/acref–9780198821472 (accessed 4 August 2021).
4. Anderson, M., (1987). The Encyclopedic Dictionary of Physical Geography. *Earth-Science Reviews*, [online] *24*(3), 231–231. Available at: https://www.sciencedirect.com/science/article/abs/pii/0012825287900389?via%3Dihub (accessed 4 August 2021).
5. Appleton, E., (1959). Dictionary of physics and allied sciences. *Journal of Atmospheric and Terrestrial Physics,* [online] *16*(3–4), 396–397. Available at: https://www.sciencedirect.com/science/article/abs/pii/0021916959900935?via%3Dihub (accessed 4 August 2021).
6. Axel, P., (1962). International Dictionary of Physics and Electronics. *American Journal of Physics,* [online] *30*(2), 150–151. Available at: https://aapt.scitation.org/doi/10.1119/1.1941949 (accessed 4 August 2021).
7. Billig, E., (1968). Encyclopaedic Dictionary of Physics: Supplementary. *Electronics and Power,* [online] *14*(7), p.299. Available at: https://digital-library.theiet.org/content/journals/10.1049/ep.1968.0259 (accessed 4 August 2021).

8. Chapple, M., (2014). *Dictionary of Physics*. [online] Available at: https://www.taylorfrancis.com/books/mono/10.4324/9781315062402/dictionary-physics-michael-chapple (accessed 4 August 2021).
9. Cox, P., (1993). *Comprehensive Dictionary of Physical Chemistry.* L. Ulicky and T.J. Kemp (editors), Ellis Horwood, Chichester, (1992). Pages 472. ISBN 0–13–151747–3. *Talanta*, [online] *40*(12), 1995–1995. Available at: https://www.sciencedirect.com/science/article/abs/pii/003991409380125B?via%3Dihub (accessed 4 August 2021).
10. Craggs, J., (1961). The International Dictionary of Physics and Electronics. *Physics Bulletin*, [online] *12*(7), 206–206. Available at: https://iopscience.iop.org/article/10.1088/0031–9112/12/7/018 (accessed 4 August 2021).
11. Craven, P., (1996). Oxford Dictionary of Physics (3rd edition). *European Journal of Physics*, [online] *17*(6). Available at: https://www.researchgate.net/publication/230913124_Oxford_Dictionary_of_Physics_3rd_edition (accessed 4 August 2021).
12. Daintith, J., (1976). *A Dictionary of Physical Sciences*. [online] 1–329. Available at: https://link.springer.com/chapter/10.1007%2F978–1–349–02886–3_1 (accessed 4 August 2021).
13. Dale, W., (1962). Encyclopedic Dictionary of Physics. *International Journal of Radiation Biology and Related Studies in Physics, Chemistry and Medicine,* [online] *5*(6), 623–623. Available at: https://www.tandfonline.com/doi/abs/10.1080/09553006214551271 (accessed 4 August 2021).
14. Dayton, I., (1964). Encyclopedic Dictionary of Physics. *American Journal of Physics*, [online] 32(8), 650–651. Available at: https://aapt.scitation.org/doi/10.1119/1.1970912 (accessed 4 August 2021).
15. Gray, H., & Brown, S., (1959). Dictionary of Physics. *Physics Today*, [online] *12*(9), 52–54. Available at: https://physicstoday.scitation.org/doi/10.1063/1.3060976 (accessed 4 August 2021).
16. Johnson, G., (2010). *Dictionary of Physics* (6th edition). pp. 123. Edited by John Daintith. Dictionary of Physics (6th edition). Oxford: Oxford University Press 2009. viii+616, ISBN: 978 0 19 923399 1 Oxford Paperback Reference. *Reference Reviews*, [online] *24*(3), 32–33. Available at: https://www.proquest.com/docview/862159142 (accessed 4 August 2021).

17. Kilby, B., (1979). A Dictionary of Physical Sciences. *Biochemical Education*, [online] 7(2), p.44. Available at: https://www.sciencedirect.com/science/article/abs/pii/0307441279901365?via%3Dihub (accessed 4 August 2021).

18. J. Thewlis, (1963). Encyclopedic Dictionary of Physics. *Nuclear Physics*, [online] *41*, 691. Available at: https://www.sciencedirect.com/science/article/abs/pii/0029558263905509?via%3Dihub (accessed 4 August 2021).

19. Lord, M., (1986). *Macmillan Dictionary of Physics.* [online] Available at: https://link.springer.com/book/10.1007%2F978–1–349–18301–2 (accessed 4 August 2021).

20. McCormick, A., (1978). A Dictionary of Physical Sciences. *Physics Bulleti*n, [online] *29*(1), 24–24. Available at: https://iopscience.iop.org/article/10.1088/0031–9112/29/1/023 (accessed 4 August 2021).

21. Michels, W. & Freeman, I., (1956). The International Dictionary of Physics and Electronics. *Physics Today,* [online] *9*(12), 28–29. Available at: https://physicstoday.scitation.org/doi/10.1063/1.3059847 (accessed 4 August 2021).

22. Moss, T., (1974). Concise Dictionary of Physics. *Infrared Physics*, [online] *14*(1), 57. Available at: https://www.sciencedirect.com/science/article/abs/pii/0020089174900086?via%3Dihub (accessed 4 August 2021).

23. O'Reilley, W., (1980). A Dictionary of Physical Sciences. *Physics of the Earth and Planetary Interiors*, [online] *23*(1), p.78. Available at: https://www.sciencedirect.com/science/article/abs/pii/0031920180900898?via%3Dihub (accessed 4 August 2021).

24. Palgrave Macmillan, (2004). *Dictionary of Physics*. [online] Available at: https://link.springer.com/book/10.1007%2F978–1–349–66022–3 (accessed 4 August 2021).

25. Pincherle, L., (1961). Encyclopedic dictionary of physics. *Solid-State Electronics*, [online] *3*(3–4), 324–325. Available at: https://www.sciencedirect.com/science/article/abs/pii/0038110161900193?via%3Dihub (accessed 4 August 2021).

26. Price, W., (1967). Encyclopedic Dictionary of Physics. *Physics Bulletin*, [online] *18*(1), 24–24. Available at: https://iopscience.iop.org/article/10.1088/0031–9112/18/1/012 (accessed 4 August 2021).

27. Putman, J., (1964). Encyclopedic dictionary of physics. *The International Journal of Applied Radiation and Isotopes,* [online] *15*(8), 510–511. Available at: https://www.sciencedirect.com/science/article/abs/pii/0020708X64901164?via%3Dihub (accessed 4 August 2021).
28. Putman, J., (1961). Encyclopedic dictionary of physics. *The International Journal of Applied Radiation and Isotopes,* [online] *12*(1–2), 70–71. Available at: https://linkinghub.elsevier.com/retrieve/pii/0020708X61900424 (accessed 4 August 2021).
29. Putman, J., (1963). Encyclopedic dictionary of physics. *The International Journal of Applied Radiation and Isotopes,* [online] *14*(8), 431–432. Available at: https://www.sciencedirect.com/science/article/abs/pii/0020708X63900279?via%3Dihub (accessed 4 August 2021).
30. Quarterly Journal of the Royal Meteorological Society, (1963). *Encyclopedic Dictionary of Physics*. Vols. 2 to 5. Edited by J. Thewlis, Oxford, London, New York (Pergamon Press), (1962). [online] *89*(380), 304–304. Available at: https://rmets.onlinelibrary.wiley.com/doi/10.1002/qj.49708938021 (accessed 4 August 2021).
31. Rennie, R., & Jonathan Law, (2019). A Dictionary of Physics. ISBN: 9780198821472
32. Stephenson, R., (1963). Encyclopedic dictionary of physics. Volumes 2 to 6 (Thewlis, J. ed.). *Journal of Chemical Education*, [online] *40*(2), p.105. Available at: https://pubs.acs.org/doi/10.1021/ed040p105.3 (accessed 4 August 2021).
33. Sykes, J., (1965). German and English Dictionary of Physics. *Nature*, [online] *207*(5000), 903–903. Available at: https://www.nature.com/articles/207903a0 (accessed 4 August 2021).
34. Symon, K., (1957). The International Dictionary of Physics and Electronics. *American Journal of Physics*, [online] *25*(6), 395–396. Available at: https://aapt.scitation.org/doi/10.1119/1.1934487 (accessed 4 August 2021).
35. Taylor, A., (1958). Dictionary of Physics. *Physics Bulletin*, [online] *9*(10), 271–271. Available at: https://iopscience.iop.org/article/10.1088/0031–9112/9/10/008 (accessed 4 August 2021).
36. Warren, J., (1986). McGraw–Hill Dictionary of Physics. *Physics Bulletin*, [online] *37*(9), 389–389. Available at: https://iopscience.iop.

org/article/10.1088/0031–9112/37/9/028 (accessed 4 August 2021).

37. Wilson, A., (1963). Encyclopedic Dictionary of Physics. *Acta Crystallographica*, [online] *16*(5), 437–437. Available at: https://onlinelibrary.wiley.com/doi/10.1107/S0365110X63001213 (accessed 4 August 2021).

38. Wood, R., (1978). Dictionary of physical metallurgy. *International Materials Reviews*, [online] *23*(1), 203–203. Available at: https://www.ingentaconnect.com/content/tandf/yimr20;jsessionid=10s344tkb664n.x-ic-live–03 (accessed 4 August 2021).

Index

P

Q

R

S

T

U

Printed in the United States
by Baker & Taylor Publisher Services